SpringerBriefs in Geoethics

Editor-in-Chief

Silvia Peppoloni, National Institute of Geophysics and Volcanology (INVG), Rome, Italy

Series Editors

Nic Bilham, University of Exeter, Penryn, UK
Peter T. Bobrowsky, Geological Survey of Canada, Sidney, Canada
Vincent S. Cronin, Baylor University, Waco, USA
Giuseppe Di Capua, National Institute of Geophysics and Volcanology (INVG), Rome, Italy
Iain Stewart, University of Plymouth, Plymouth, UK
Artur Sá, University of Trás-os-Montes and Alto Douro, Vila Real, Portugal
Rika Preiser, Stellenbosch University, Stellenbosch, South Africa

SpringerBriefs in Geoethics envisions a series of short publications that aim to discuss ethical, social, and cultural implications of geosciences knowledge, education, research, practice and communication. The series SpringerBriefs in Geoethics is sponsored by the IAPG—International Association for Promoting Geoethics (http://www.geoethics.org).

The intention is to present concise summaries of cutting-edge theoretical aspects, research, practical applications, as well as case-studies across a wide spectrum.

SpringerBriefs in Geoethics are seen as complementing monographs and journal articles, or developing innovative perspectives with compact volumes of 50 to 125 pages, covering a wide range of contents comprising philosophy of geosciences and history of geosciences thinking; research integrity and professionalism in geosciences; working climate issues and related aspects; geoethics in georisks and disaster risk reduction; responsible georesources management; ethical and social aspects in geoeducation and geosciences communication; geoethics applied to different geoscience fields including economic geology, paleontology, forensic geology and medical geology; ethical and societal relevance of geoheritage and geodiversity; sociological aspects in geosciences and geosciences-society-policy interface; geosciences for sustainable and responsible development; geoethical implications in global and local changes of socio-ecological systems; ethics in geoengineering; ethical issues in climate change and ocean science studies; ethical implications in geosciences data life cycle and big data; ethical and social matters in the international geoscience cooperation.

Typical topics might include:

- Presentations of core concepts
- Timely reports on state-of-the art
- Bridges between new research results and contextual literature reviews
- Innovative and original perspectives
- Snapshots of hot or emerging topics
- In-depth case studies or examples

All projects will be submitted to the series-editor for consideration and editorial review.

Each volume is copyrighted in the name of the authors. The authors and IAPG retain the right to post a pre-publication version on their respective websites.

The Series in Geoethics is initiated and supervised by Silvia Peppoloni and an editorial board formed by Nic Bilham, Peter T. Bobrowsky, Vincent S. Cronin, Giuseppe Di Capua, Rika Preiser, Artur Agostinho de Abreu e Sá, Iain Stewart.

More information about this series at http://www.springer.com/series/16482

Jan Boon

Relationships and the Course of Social Events During Mineral Exploration

An Applied Sociology Approach

 Springer

Jan Boon
FaciliTech International
Ottawa, ON, Canada

ISSN 2662-6780 ISSN 2662-6799 (electronic)
SpringerBriefs in Geoethics
ISBN 978-3-030-37925-4 ISBN 978-3-030-37926-1 (eBook)
https://doi.org/10.1007/978-3-030-37926-1

This Springer imprint is published by the registered company Springer Nature Switzerland AG
The registered company address is: Gewerbestrasse 11, 6330 Cham, Switzerland

IAPG Foreword

Living sustainably, prosperously and equitably on our crowded planet in the coming decades will depend on mining. However rapidly we increase recycling rates, improve resource efficiency and reduce demand for raw materials through new approaches to product design and use, we will continue to need to mine significant quantities of an ever-increasing range of elements. The mineral needs of the near future will be quite different to those of the recent past, given the urgent need to transition to low-carbon energy systems and to harness new, materially complex technologies to address a nexus of environmental, social and economic challenges, as articulated in the UN Sustainable Development Goals. Meeting these needs will mean mining in new places and communities—as well as in settings that bear the scars of unethical and unsustainable practices of the past—and will depend on the engagement and support of communities rightly seeking to assert their rights and defend their interests. It is therefore essential, from both a moral and practical standpoint, to mine responsibly, minimizing negative social and environmental impacts, maximizing benefits and legacies to affected communities, and including them as partners in a shared societal enterprise.

Finding and winning resources from the Earth's crust to meet global demand, while also addressing the needs and interests of local communities and protecting ecosystems, undoubtedly requires "research and reflection on the values which underpin appropriate behaviours and practices, wherever human activities interact with the Earth system"—the definition of geoethics offered by the 2016 Cape Town Statement on Geoethics. Furthermore, while the principal focus of geoethics was initially on the activities of professional geoscientists, its practical scope has broadened to encompass a wide range of other human agents. Mining in the twenty-first century exemplifies the entanglement of geoscientists' roles and activities with those of other professionals and multiple overlapping non-specialist groups (including consumers, company employees and local communities), and natural systems—and therefore sits at the forefront of the burgeoning field of geoethics.

It is therefore particularly apposite that the first in this series of *SpringerBriefs in Geoethics* should be on this topic. It is also fitting that it should typify two other current key trends in geoethics. First, it radically crosses disciplinary boundaries and communities of practice. Its author, drawing on his varied and distinguished career in the natural resources sector, views a recognized challenge in that sector—the need to establish and maintain effective relationships with stakeholders and communities—through the novel lens of sociological theory. In doing so, he makes a compelling case for the value of bringing such unfamiliar intellectual tools (applied with appropriate methodological rigour) to bear on familiar problems. Second, the rich insights that he draws from his analysis span theory and practice. They elucidate and enhance the value of the sociological theory of symbolic interactionism in a novel setting, add to the theoretical "toolkit" of geoethics and provide clear, practical lessons for policy-makers, mining companies and local communities, for the benefit of people and planet.

November 2019

Nic Bilham
University of Exeter
Exeter, UK

Giuseppe Di Capua
Istituto Nazionale di Geofisica
e Vulcanologia
Rome, Italy

International Association
for Promoting Geoethics

Preface

While most social responsibility codes and guidelines for the mineral exploration and mining industry emphasize the importance of establishing and maintaining productive relationships between relevant actors, they do not pay specific attention to the sociological processes that underlie these relationships. I undertook my Ph.D. thesis project to shine light on these processes, and I base this book on my thesis "Corporate Social Responsibility, Relationships and the Course of Events in Mineral Exploration—an Exploratory Study" (Boon, 2015).

This book's focus on the human side of mineral exploration supports the part of the Cape Town Statement on Geoethics that says, "Geoscientists have specific knowledge and skills, which are required to... support human life and well-being ... and to ensure natural resources are managed and used sustainably. This entails ethical obligations. Therefore, geoscientists must embrace ethical values in order best to serve the public good." (Cape Town Statement on Geoethics 2016).

Ottawa, Canada   Jan Boon

Reference

Boon, J. (2015). *Corporate social responsibility, relationships and the course of events in mineral exploration—An exploratory study* (Unpublished Ph.D. thesis). Carleton University, Ottawa. https://curve.carleton.ca/system/files/etd/6c6598d4-c436-409e-9ba1-40dea2d37d2c/etd_pdf/7b39ca613ff7e7e2df52ed82580e3974/boon-corporatesocialresponsibilityrelationships.pdf.

Acknowledgements

I very much appreciate my wife's encouragement, her acceptance of my frequent absences for fieldwork and her patience in living with a person seemingly forever buried in studies. In addition, the project would not have been possible without the help and support of many people, institutions and organizations. I am grateful to all those who agreed to be interviewed and gave me their time and shared their experiences with me.

I received financial support from Ontario Graduate Scholarships, the University of Ottawa, the Organization of American States (OAS) and a 50/50 shared grant from the Prospectors and Developers Association of Canada (PDAC) and MITACS (a federal–provincial granting agency). The Secretaría Nacional de Educación Superior, Ciencia, Tecnología e Innovación del Ecuador (SENESCYT) through its programme "Prometeo—Viejos Sabios" provided financial support for a stay in Ecuador to develop the social responsibility strategy of the Instituto Nacional de Investigación Geológico, Minero y Metalúrgico del Ecuador (INIGEMM), and to conduct the Ecuadorian case studies. INIGEMM provided logistics and staff support.

Dr. Wallace Clement supervised the Ph.D. project at Carleton University together with the members of the internal thesis committee: Dr. Neil Gerlach and Dr. Sefa Hayibor. Lic. Roberto Sarudiansky of the Universidad Nacional de San Martín (UNSAM) in Buenos Aires was co-supervisor for the OAS-funded parts of the study.

In Ecuador, Dominic Channer and María Clara Herdoíza of Kinross/Aurelian Ecuador and their community relations team (Fruta del Norte) and Jorge Barreno and Fernando Carrión of INV Metals and their community relations team (Loma Larga) allowed us to undertake our studies, helped setting up the interviews and provided logistic support. Jéssica Marcayata and Dayana Velasco of the INIGEMM participated in the interviews of the Fruta del Norte study together with me, and Camilo Aguas safely transported us there and back. The Social Responsibility Committee of the INIGEMM helped with development of the questionnaire and gave moral encouragement. Colón Velásquez, the Executive Director of the

INIGEMM, provided me with an office and support, and approved the assignment of personnel and resources to the field studies. I shared an office with Michel Rueda, who helped me in an endless numbers of ways. Ricardo Valdez of the Canadian embassy in Quito provided invaluable assistance in overcoming hurdles in the bureaucratic processes related to my travel to Ecuador, helped me find a replacement when one project fell through and introduced me to many people in the mineral exploration and mining sector of Ecuador.

Juan José Herrera and Felipe Injoque of Volcan Compañía Minera approved the case study of the Palma project, and Luis Rojas travelled with me up and down the Lurín valley to conduct interviews, with Yanina and Luis as our able drivers. Oscar Vásquez, Nuala Lawlor and Ariana Vindrola of the Canadian embassy in Lima were always there when I needed them.

In Argentina, Waldo Pérez of Lithium Americas enthusiastically supported our study of the Cauchari-Olaroz project, and we received field support from Santiago Campellone of Minera Exar and his community relations team Mónica Echenique and Gilda Agostino. Graciela Medardi of the Universidad Nacional de Jujuy helped us identify non-company interviewees and set up the interviews that Bernarda Elizalde and I conducted. Facundo Huidobro of Mansfield Minera introduced me to the members of the Salta Chamber of Mines, and together with Mabel Casimiro, helped me set up and conduct interviews. Ana Garasino of the Canadian embassy helped me in more ways than I can mention here. She helped me gain an understanding of the Argentine context for mineral exploration and mining and gave me access to the mineral exploration and mining community. Pablo Lumerman of Estudios del Valle kindly looked at the flow of the narrative.

My colleagues at Natural Resources Canada supported me throughout, and I am especially grateful to Lise-Aurore Lapalme who meticulously commented on every draft chapter of my thesis.

Contents

Acronyms

APEOSAE	Asociación de Pequeños Exportadores Agropecuarios Orgánicos del Sur de la Amazonía Ecuatoriana (Association of Small Organic Farming Exporters of Southern Ecuadorian Amazonia)
C.C.	Comunidad Campesina (Peasant Community [legally defined entity])
C.P.	Centro Poblado (Populated settlement [legally defined subset of a C.C.])
CIM	Canadian Institute of Mining
COFENAC	Consejo Cafetalero Nacional (National Coffee Council [Ecuador])
CONAIE	Confederación de Naciones Indígenas del Ecuador (Confederation of Indigenous Nations of Ecuador)
CSR	Corporate Social Responsibility
CSV	Creating Shared Value
DFATD	Department of Foreign Affairs, Trade and Development (Government of Canada. Now renamed as Global Affairs Canada)
ECUARUNARI	Ecuador Runakunapak Rikcharimuy (Confederation of Kichwa Peoples of Ecuador)
EIA	Environmental impact assessment
ETAPA	Empresa Pública Municipal de Teléfonos, Agua Potable y Alcantarillado (Public Municipal Telephone, Drinking Water and Sewage Systems Company [of the city of Cuenca, Ecuador])
FOA	Federación de Organizaciones Indígenas y Campesinas del Azuay (Federation of Indigenous and Farmer Organizations of Azuay [Ecuador])
GAD	Gobierno Autónomo Descentralizado (Autonomous Decentralized Government [Ecuador])

IIMP   Instituto de Ingenieros de Minas del Perú (Peruvian Institute of
                Mining Engineers)
IMF   International Monetary Fund
INIGEMM   Instituto Nacional de Investigación Geológico, Minero y
                Metalúrgico del Ecuador (National Institute for Geological,
                Mining and Metallurgical Research of Ecuador)
MITACS   Mathematics of Information Technology and Complex Systems
                (definition appears rarely) (Federal–provincial granting agency
                in Canada)
NCP   National Contact Point for the Organization of Economic
                Cooperation and Development dispute resolution mechanism
NGO   Non-governmental organization
OAS   Organization of American States
OECD   Organization of Economic Cooperation and Development
OGS   Ontario Graduate Scholarships
ONDS   Oficina Nacional de Diálogo y Sostenibilidad (National
                Dialogue and Sustainability Office [Peru])
PDAC   Prospectors and Developers Association of Canada
PDYOTLE   Plan de Desarrollo y Ordenamiento Territorial de Los
                Encuentros (Los Encuentros Development and Land Use Plan
                [Ecuador])
PRONOEI   Programa No Escolarizado de Educación Inicial
                (Community-based early education programme [Peru])
SENESCYT   Secretaría Nacional de Educación Superior, Ciencia, Tecnología
                e Innovación del Ecuador (National Secretariat for Higher
                Education, Science, Technology and Innovation of Ecuador)
SENPLADES   Secretaría Nacional de Planificación y Desarrollo del Ecuador
                (National Secretariat for Planning and Development of Ecuador)
UNAGUA   Unión de Sistemas Comunitarios de Agua del Azuay (Union of
                Community Water Systems of Azuay [Ecuador])
UNSAM   Universidad Nacional de San Martín (National University of
                San Martín [Argentina])

List of Figures

List of Tables

Chapter 1
Introduction

In 2013, over 1000 Canadian mineral exploration companies were active in more than 130 countries, often in remote locations (Drake, 2013), and companies listed on Canadian stock exchanges raised 40% of equity financing for mineral exploration in 2011 worldwide (Natural Resources Canada, 2013). Among the many issues that arise in their interaction with local communities, conflicts and the relative distribution of benefits and harms are prominent. Many international institutions and industry organizations have developed guidelines to help address those issues. While they mention the importance of establishing and maintaining productive relationships between relevant actors, they lack a sociological framework for understanding these relationships and the processes through which they exert their influence. My Ph.D. thesis project aimed to fill this gap.

I conducted case studies of nine mineral exploration projects: one in Canada, one in Mexico, three in Ecuador, two in Peru and two in Argentina. Because of space limitations, I will describe only five of these case studies in this book. My Ph.D. thesis provides details for all case studies (Boon, 2015). The theory of symbolic interactionism, together with relationship indicators developed for the purpose of this study, formed the basis for analysing and interpreting the results and for developing a generalized model of the processes that drive the course of social events and the perceptions of benefits and harms. This approach presents a coherent framework for understanding the interactions and relationships between the actors, the underlying processes and the influence of these processes on the course of social events and on the perceptions of present and future benefits and harms in mineral exploration projects.

Chapter 2 provides the context in which mineral exploration companies operate. Their activities focus on the early part of the mining cycle. The revenue of companies involved in the stages up to exploitation derives from the sale of the rights to develop the prospect to the entity that operates the next stage. On average, only a low proportion of exploration projects ever become a mine and the high financial risk involved at most stages creates considerable uncertainty, which in turn affects community relations. Maintaining conflict-free socially responsible operations in a globalized environment in which projects are under constant scrutiny is a challenge,

© The Author(s), under exclusive license to Springer Nature Switzerland AG 2020 1
J. Boon, *Relationships and the Course of Social Events During Mineral Exploration*,
SpringerBriefs in Geoethics, https://doi.org/10.1007/978-3-030-37926-1_1

and the industry is resorting to Corporate Social Responsibility (CSR) strategies to try to meet it. Mineral exploration companies tend to be small, operationally nimble, risk-taking and adventurous and perennially on the lookout for capital to finance their operations.

The actor groups that play a role in mineral exploration include companies, non-government organizations (NGOs), governments at all levels, communities, investors and financiers. Chapter 2 briefly outlines the characteristics of each.

As social conflicts that arise during the exploration stage often affect the course of social events during the subsequent stages of the mining cycle, it is important to understand and, where possible, address them at this stage. Factors that play a role in the occurrence or absence of social conflicts include cultural differences, local and country history, socio-economic conditions, development models, and company and community characteristics and actions. Social conflicts occur in all industrial sectors and are the subject of increasing attention, which has led to the phenomenal growth of CSR over the past few decades. There is a great variety of definitions of the CSR concept, almost all of which view it as a voluntary effort over and above the requirements of the law. The simplest definition is that of the European Commission: "Corporate Social Responsibility is the responsibility of enterprises for their impacts on society" (European Commission, 2011). Chapter 2 provides a brief update on the field and suggests that CSR as currently practised does not appear to be the sole answer to the problems it hopes to solve. Few companies, if any, look at larger systemic issues such as the structure of the industry or at alternative economic models. Many communities have little understanding of the CSR concept, but many view it as the company's duty to make considerable (mainly financial) contributions, while many host and home governments see CSR as a development tool. While they can contribute to development and often do, this is not their mission. CSR initiatives have had mixed results, and a broader, more transformational change may be needed (Bowen, Newenham-Kahindi, & Herremans, 2010). Chapter 2 considers the adjective "corporate" to be misleading as it draws attention away from the fact that all actors involved need to be socially responsible. The characteristics of the transformational approaches described by Bowen et al., collaborative processes, community decision-making, acting together, supporting, empowerment, leadership and intensive alliances, lend support to this contention. Unfortunately, these characteristics are often lacking. Chapters 6 and 7 show that application of the model developed as part of the present study helps promote these characteristics, and in addition, helps meet the expectations referred to above.

While there has been long-term recognition of the importance of relationships in mineral exploration, and while stakeholder approaches are quite common, the sociological processes that play a role in relationships and interactions, and their effects have received little attention. This led me to focus on the processes through which relationships influence the course of events, and perceived benefits and harms related to mineral exploration projects.

Chapter 3 discusses symbolic interaction theory and identifies meanings; relationships and associated interactions; interpretations; decisions; and reference communities as the key concepts involved in the micro-processes at work. It discusses each of these concepts in some detail together with field observations from the present study. In addition, it explains and stresses the importance of meeting transactional needs (identity verification; benefits from the encounter; inclusion; trust; transparency) in interactions, together with the advisability of using the tactical dimensions of building trust (visibility; sincerity and personalization; showing face; and establishing routines). As individuals have multiple identities, each of which carries its own network of relationships and interactions, and each of which draws on a different reference community, situations can be quite complex. This chapter also explains how group identities form and argues that the interplay between the various identities could be a driver of change. It further shows the link between group identity and interpersonal networks, drawing on the work of Deaux and Martin (2003).

As interactions and relationships continually adjust meanings, interpretations and decisions, and as relationships strongly influence interactions, I developed a tool for characterizing relationships. The tool describes relationships in terms of the indicators trust; respect; communication; mutual understanding, conflict resolution; goal compatibility; balance of power; focus; frequency; stability; and productivity.

Chapter 4 provides a brief outline of the research methods employed and describes each case study using the headings: field study information; project description; context; patterns and characteristics of relationships; company social responsibility approach; and perceived present and future benefits and harms.

Chapter 5 analyses and interprets the field study results. For each case study, I was able to assign a risk-of-conflict measure to each relationship indicator, which resulted in a "relationship indicator profile". Combination of these profiles into a comparison table clearly differentiated the cases. An interpretation of the field observations in terms of symbolic interactionist processes involving meanings, reference communities, relationship patterns and change showed that the relationship indicators coupled with a symbolic interactionist framework successfully linked micro-level interactions to meso-level actions and provided a credible explanation of the factors influencing the course of social events surrounding mineral exploration projects.

Symbolic interactionism provided an excellent theoretical and methodological basis for describing and understanding the processes that link relationships to the course of social events and to perceived harms and benefits. It did so by paying close attention to interactions at the person-to-person level through the concept of meanings, then tracing the path to decisions weighed against reference community norms and following the change process by identifying the nodes where the change of meanings begins and how it diffuses out into the community. The relationship patterns formed by repeated cycling through this process lead to new social structures. The relationship characterization tool allowed characterization of relationship patterns in a relatively simple way.

Chapter 6 presents a generalized interactionist model for mineral exploration projects that links relationships to the course of social events and to perceived present and future benefits and harms associated with a mineral exploration project. The model consists of seven "core stages" with a number of additional influences feeding into it. The chapter explains each of these stages and influences, and their link to the course of social events and perceptions of present and future benefits and harms.

Chapter 7 considers the implications of my findings and summarizes my conclusions. Suggestions applicable to all actors relate to reference communities and meanings; the importance of self-analysis and dialogue; and relationship development. The chapter makes additional specific suggestions for home and host governments, industry, and communities.

References

Boon, J. (2015). *Corporate social responsibility, relationships and the course of events in mineral exploration—An exploratory study* (Unpublished Ph.D. thesis). Carleton University, Ottawa. https://curve.carleton.ca/system/files/etd/6c6598d4-c436-409e-9ba1-40dea2d37d2c/etd_pdf/ 7b39ca613ff7e7e2df52ed82580e3974/boon-corporatesocialresponsibilityrelationships.pdf.

Bowen, F., Newenham-Kahindi, A., & Herremans, I. (2010). When suits meet roots: The antecedents and consequences of community engagement strategy. *Journal of Business Ethics, 95*(2), 297–318.

Deaux, K., & Martin, D. (2003). Interpersonal networks and social categories: Specifying levels of context in identity processes. *Social Psychology Quarterly, 66*(2), 101–117.

Drake, A. (2013). Canadian global exploration activity. Retrieved July 30, 2014 from http://www.nrcan.gc.ca/mining-materials/exploration/8296.

European Commission. (2011). *A renewed EU strategy 2011–14 for corporate social responsibility* (Communication from the Commission to the European Parliament, the Council, the European Economic and Social Committee and the Committee of the Regions No. COM (2011) 681 final). Brussels: European Commission.

Natural Resources Canada. (2013). Canada is a global mineral exploration and mining giant. Retrieved July 30, 2014 from http://www.nrcan.gc.ca/mining-materials/exploration/8296.

Chapter 2
Context

This chapter summarizes the context in which mineral exploration takes place. It discusses the mining cycle, describes the groups of actors involved in or affected by mineral exploration and looks at the risk of social conflict. In addition, it describes the social responsibility concept, its evolution and related international codes and guidelines. Finally, it pays attention to the special place occupied by indigenous people and the concept of Free, Prior and Informed Consent (or Consultation) and comments on the distance that still needs to be covered.

2.1 The Mining Cycle

The mining cycle starts with the collection and interpretation of basic geoscience data by government geological surveys. Prospectors use this information and their own knowledge and intuition to identify promising areas and stake claims. Junior companies then buy the rights to these claims and raise funds to conduct early stage exploration (geophysical and geochemical surveys, drilling). If their work indicates significant resource potential, they sell the rights to their claims to larger companies that conduct advanced exploration (detailed surveys and drilling campaigns) and development (infrastructure, economic feasibility studies and environmental impact assessments). If results continue to be encouraging and investment funds are available, the construction of a mine can begin. Only when production starts will the project generate its first income. Once exploitation exhausts the ore body (which can take up to 30 years or longer), the mine needs to be closed, meeting a series of environmental, safety and security requirements. The social issues that may arise vary from stage to stage of the cycle.

The relative simplicity of the operations of geoscience information collection and prospecting should reduce the probability of social conflict at these stages. However, this is not always the case; for example, geologists of the Instituto Nacional de Investigación Geológico, Minero y Metalúrgico del Ecuador (INIGEMM) reported that they have had to flee the scene on some occasions to avoid physical harm.

J. Boon, *Relationships and the Course of Social Events During Mineral Exploration*, SpringerBriefs in Geoethics, https://doi.org/10.1007/978-3-030-37926-1_2

Small companies, many of whom lack the skills needed for community engagement, usually carry out early stage exploration and therefore they may, albeit not intentionally, cause friction and suspicion. On the other hand, their small size means that once they find ways to implement socially responsible practices, they can do it quickly and with good internal buy-in. The advanced exploration stage involves diamond drills, possibly earth moving equipment and crews that can have up to 50 members. Because of their greater impact, social aspects demand greater care. The probability that an exploration target becomes a mine is low. In an area where mines are already operating it is about 1 in 24, while in greenfield exploration it is between 1 in 1000 and 1 in 3333 (Kreuzer & Etheridge, 2010). Communities do not know this and may have unrealistic expectations, both about the prospects of a prosperous future and about the "depth of the junior company's pockets". In addition, activities during the exploration stage can vary strongly with time, for example, when a period of drill core analysis in the laboratory and there is no field activity follows, a period of extensive drilling. This often confuses communities that sometimes are not sure whether or not a project is still ongoing or whether equipment left behind will stay there forever. Therefore, managing expectations during the exploration stages are very important.

The mine construction phase involves contractors that are only on-site for a short time. They may not pay much attention to local social issues, which, together with the general disturbance created by construction activities, increases the risk of social conflict.

The operations and closure phases each involve social and environmental risks that are different from those of the preceding parts of the mining cycle. Closure requires careful planning that should start at the beginning of operations.

The total time elapsed between geoscience studies and completion of mining infrastructure development can be anywhere from five to 20 years. Furthermore, total investment up to the start of the mining stage can amount to billions of dollars. Issues arising during these stages may relate to access to land; environmental protection; local and national power relations; employment; human rights; safety and security; social impact; and distribution of risks and benefits. Communities see the entire cycle as a single undertaking and do not distinguish between the different actors involved in the various stages. Therefore, events during the early stages can have a profound effect on later stages and it is important to understand the social processes that occur during the exploration stage because of their potentially lasting effect.

2.2 Conflicts

Over the past decade and a half, communities have steadily gained power through instruments such as ILO 169 (International Labour Organization, 2012), the United Nations Declaration of the Rights of Indigenous Peoples, their right to Free, Prior and Informed Consent (or Consultation) (United Nations, 2007), and in Canada the Crown's duty to consult. The industry is well aware that, even though the authorities may have approved their project, developing constructive relations with the communities affected by their activities is essential for their success. This requirement has come to be called "social licence to operate", a concept criticized by many. According to Boutilier and Thomson (2011), "social licence to operate" is a process, not a transaction. Unfortunately, the word "licence" suggests an instrumental, purely transactional mechanism, and many in the industry treat it as such.

Distinguishing features of mineral exploration include small company size, therefore, less resources and maybe less power. In addition, limited resources and investor focus on quick returns may make it difficult to add social expertise to the staff complement. However, investor attitudes are rapidly changing and many now include social considerations into their due diligence. While exploration has much less environmental impact than mining, there is a risk of unethical behaviour: many juniors are in the business to prove up a prospect and then sell it to a medium-size or large mining company. As the selling price depends on the perceived mineral potential, there is a strong temptation to exaggerate findings. The well-known Bre-X scandal is an extreme example, which led stock exchanges to tighten regulations (Ankli & Varadan, 1999).

The large number of exploration companies makes it difficult to make voluntary industry standards binding (Gunningham & Rees, 1997) and the industry's trade organization (Prospectors and Developers Association of Canada) has not (yet?) made adhering to its voluntary social responsibility framework a condition of membership. The power imbalance between exploration companies and communities, while it can exist, is usually less than in mining. Communities can use the ability to withhold approval, to organize protests and to disrupt operations as power tools more effectively against a smaller company. In cases where communities have well-defined rights to the land there can be power equilibrium. As I will show later, this was the case in the Cauchari-Olaroz case study.

Actor groups related to mineral exploration and mining include corporations, NGOs, local and national governments, international institutions, communities and investors.

The powers at the disposal of governments vary with circumstances and countries. The governments of developed nations helped set the agenda and a context for corporate behaviour through their influence in international bodies and the means at their disposal to influence the foreign activities of companies headquartered in their countries (e.g. diplomatic assistance, tax rules, capacity building and public shaming). The powers of governments of developing countries are limited in a number of ways. First, their need for foreign direct investment to create jobs and generate income puts them in a weak negotiating position. Also, state dependence upon loans from institutions such as the World Bank or the IMF, whose loan conditions required export-led industrialization, often has provided incentives to promote industrial expansion at the expense of social and environmental safeguards, which sometimes led to communities becoming the victims of state aggression or discrimination (Garvey & Newell, 2005). Second, national and local governments often lack resources and capability and are poorly prepared for effective cooperation with both communities and corporations in development programs, and they frequently lack the resources to enforce regulations (International Development Research Centre, 2003). Third, many governments, especially in developing nations face the challenge of bribery and corruption. The Extractive Industries Transparency Initiative and the OECD Guidelines for Multinational Enterprises (Extractive Industries Transparency Initiative; OECD, 2011) contain clauses that specifically aim to reduce the incidence of such situations. In Canada, the Government of Canada updated the Corruption of Foreign Public Officials Act in 2013 and considerably increased its enforcement activities (Hutton & Beaudry, 2014). It also has mandated disclosure of payments to all levels of government through the Extractive Sector Transparency Measures Act that came into force in June 2015 (Natural Resources Canada, 2015). Finally, groups or constituencies that are more powerful receive government support while communities do not (Garvey & Newell, 2005: 393).

Extractive industries have to be located where the resource is, often far removed from the centres of political power and mainstream economic activity. Communities are heterogeneous and consist of subgroups categorized by income, age, gender, occupation, religion, education level, part of town, dialect and others. A single individual can have multiple identities related to his or her membership of different subgroups. As I will demonstrate later, interpersonal relationships play an important role. The internal dynamics of a community can be quite complex and influence how individual choices are converted into community choices, a process that affects the future of both the community and the corporation that plans to establish operations there (Garvey & Newell, 2005: 399, 400). The sociological approach that I will discuss later provides a useful framework for understanding community internal dynamics related to mineral exploration projects. Mineral exploration projects affect communities within their zone of influence in a number of ways. Not only does the company occupy the physical environment with its equipment, buildings and other infrastructure but also its economic weight, employment practices, work organization and hierarchies, and value system inevitably influence the social structures, values, worldview and attitudes of the communities, especially when the latter are small. At the same time, the community influences the company and its internal culture, as I will show. When communities need to arrive at a decision on a proposed corporate development, negotiate a deal and ensure that, once in operation, the corporation will adhere to its commitments, they face a number of issues. First, during the proposal stage, who will guide the debate and how? How to ensure representation of all views in the community? How to decide when a position is ready to go forward? How to maintain unity? While ultimately the community has to make its own decisions, NGOs and unions can provide useful assistance, provided it is possible to address any questions about their own accountability. Companies may just want to "buy peace" during critical initial project phases and focus on selective parts of the communities that are most likely to be affected by these initial project operations, and any social support initiatives may not be sustained once the critical project phases have been completed (Frynas, 2005: 585). Communities must also find ways to understand the technical aspects and potential impacts of the proposed development. In line with the endogenous development paradigm, it may be advantageous for communities to work with local universities or professional organizations where possible (Mas Herrera, 2005).

A host of basic themes can lead to social conflict. Among the most important are the environment, water, competition for resources and development models. de Echave et al. presented a more detailed list (de Echave, Diez, Huber, Ricard Lanatta, & Tanaka, 2009).

2.3 Social Responsibility

As the extractive industries view community relations through a Corporate Social Responsibility lens, and as establishing relationships with communities is a key component of CSR strategies, an understanding of the concept and its evolution provides a context for the present book. This section discusses social responsibility in a broad sense. The mineral exploration and mining industry has embraced the concept of Corporate Social Responsibility (CSR) and has adopted a discourse of "socially and environmentally responsible mining", or "new mining" (Salas Carreño, 2008). This discourse is permeating the industry, together with its associated guidelines, codes and private regulation frameworks. While many view it more as a "technique" than as a foundational philosophy, some have linked it to deeper philosophical considerations. For example, Humberto Ortiz Roca put it in the ethical context of the Social Doctrine of the Catholic Church (Ortiz Roca, 2012).

The past decades have seen the development of a plethora of principles, guides, codes, reporting guidelines and similar documents related to CSR in mining and mineral exploration. Well-known approaches include the e3 Plus Framework for Responsible Exploration of the Prospectors and Developers Association of Canada (PDAC), Towards Sustainable Mining of the Mining Association of Canada (MAC) and the International Finance Corporation Performance Standards. In addition, there are the Global Reporting Initiative (that began in 1999 and rapidly gained strength over the past decade) and the Initiative for Responsible Mining Assurance. The Responsible Mining Index and ISO 26000 close this incomplete list (International Finance Corporation (IFC), 2012; ISO, 2010; Mining Association of Canada, 2010; Prospectors and Developers Association of Canada, 2010; Initiative for Responsible Mining Assurance, 2017; Responsible Mining Index, 2017). Note that ISO 26000 is not a certifiable standard. In addition to firms specializing in CSR, many existing legal, organizational development and management-consulting firms are adding a CSR component to their offerings, and more and more companies publish CSR statements, CSR protocols and produce "CSR", "Social" or "Sustainability" reports. Impressive machinery has come into being, most of it instrumental in its approach. In June 2011, the United Nations Human Rights Council unanimously endorsed the final report of John Ruggie, the Secretary General's Special Representative on Issues of Human Rights, Transnational Corporations, Other Business Enterprises: "Guiding Principles on Business and Human Rights: Implementing the United Nations "Protect, Respect and Remedy" Framework" (Ruggie, 2011). The Framework itself, published in 2008, addresses the state's duty to protect against human rights abuses, the corporate responsibility to respect human rights and the responsibility of both the state and corporations to provide access to remedies (Ruggie, 2008). The "Ruggie reports" are playing a very influential role in the development of human rights practices all over the world and have become part of both the International Finance Corporation Performance Standards and the Organization for Economic Cooperation and Development (OECD) Guidelines for Multinational Enterprises (International Finance Corporation (IFC), 2012; OECD, 2011). Vanclay and Hanna recently provided an up to date summary of codes and guidelines (2019).

These codes and guidelines have a number of elements in common that include transparent and ethical behaviour; compliance with applicable law; consistency with international norms of behaviour; taking into account the expectations of stakeholders; integration throughout the organization; and practicing these in the organization's relationships. For geoscientists working in mineral exploration, this includes Geoethics.

A number of alternative concepts contain elements that the reader may find useful. They include:

- The Social and Environmental Value Governance Ecosystem (SEVGE) approach views corporations as only one of many actors in a complex ecosystem with environmental and social values at its core and overall system governance taking place through a variety of mechanisms, one of which is CSR (Sagebien & Lindsay, 2011). It sees corruption as a pervasive disabling factor, whereas it views dialogue and CSR as enabling factors.
- Creating Shared Value (CSV) is defined as "policies and operating practices that enhance the competitiveness of a company while simultaneously advancing the economic and social conditions in the communities in which it operates". It focuses on identifying and expanding the connections between societal and economic progress (Porter, 2011; Porter & Kramer, 2011).
- "Integral Social Responsibility" that applies to all actors (Guay, 2012). According to this model, governments and communities as well as companies have social responsibilities.
- The Social Doctrine of the Catholic Church puts humanity rather than the firm at the centre. It assigns corresponding responsibilities to businesses and their owners and managers and contains extensive sections on human rights and the environment (Pontifical Council for Justice and Peace, 2005).

Aboriginal communities occupy a special place among communities, as two important documents govern relations with them: the 1989 International Labour Organization Convention 169, commonly referred to as "ILO 169", and the 2007 United Nations Declaration on the Rights of Indigenous Peoples (International Labour Organization, 2008, 2012; United Nations, 2007). ILO 169 recognizes land and property rights, equality and liberty and autonomy for indigenous peoples. The UN declaration emphasizes, among other things, the freedom from discrimination, the right to self-determination, self-government and the right to live in freedom, peace and security as distinct peoples. In addition, it asserts the right not to suffer forced assimilation, states that indigenous peoples shall not be forcibly removed from their land or territory—this includes the right to Free, Prior and Informed Consent (FPIC) and the right to participate in decision-making that affects indigenous rights. States are obliged to establish mechanisms to safeguard these rights. Section 35 of Canada's Constitution Act 1982 recognizes and affirms the existing aboriginal and treaty rights of aboriginals. A number of Supreme Court of Canada rulings oblige governments to consult the potentially affected aboriginal communities when they make decisions that may have an impact on aboriginal rights or treaty rights (even prior to final proof of the rights in court or final settlement on the rights in negotiation processes) (Department of Justice, 2012; Newman, 2009). A number of industry associations have established aboriginal affairs committees that have led to aboriginal people assuming a higher profile in the industry. For example, a past president of the PDAC is aboriginal, aboriginal people are members of the boards of various Canadian companies, and First Nations communities are becoming participants in mineral exploration and mining ventures. Agreements between industry organizations and aboriginal or community organizations are also becoming more common. Examples include the Memorandums of Understanding between the Canadian Assembly of First Nations (AFN) and the PDAC and the Mining Association of Canada (Mining Works for Canada, 2009; Prospectors and Developers Association of Canada, 2008). This is also happening in other countries, for example, the Chamber of Mining of the province of Salta in Argentina has signed an agreement with the Colla aboriginal communities of Salta (Gobierno de Salta, 2012). The aboriginal communities of the Departamento de los Andes of the Province of Jujuy (Argentina) published a declaration in which they supported mining. They emphasized that it should be done in a sustainable fashion and that industry should provide employment and help build capacity in the communities (Conciencia Minera, 2012), and a number of communities in the Department of Susques pushed Jujuy provincial legislators to approve the environmental impact study of Minera Exar (personal communication, Santiago Campellone).

There is currently much debate in Canada about benefit-sharing models as a focus of approaches to community development agreements. These include royalty and profit sharing, equity stakes, community foundations and assigning part of government revenues to local development initiatives. Impact Benefits Agreements (IBAs) have been part of the standard approach in Canada for the past two decades and many have been quite successful (IBA Research Network, 2010).

While much distance remains to be covered, the mineral exploration and mining industry, in general, has made important progress in how it deals with community-related matters within existing systemic conditions.

There is a significant risk that the gradual movement of part of the industry towards new vistas will be more than offset by the practices of part of the industry that is using CSR as a fig leaf. When trying to bring business (back?) in line with society, companies have to demonstrate that they walk the talk. Any serious misbehaviour will create public cynicism and mistrust not only of the particular company that is guilty of it, but by association of the entire sector. Walking the talk requires companies to embed CSR in all organizational behaviours and actions, a goal that is difficult to achieve for many reasons. Gunningham and Rees argued that there is a need for an effective industrial morality, and they described conditions under which such a morality could form, many of which involve a strong role for governments. They emphasized that mechanisms for dealing with "bad apples" are crucial to success. Their work makes it clear that this is a major undertaking with no assurance of success (Gunningham & Rees, 1997).

Breaking the mould requires transformative change, and experimentation in social enterprises that are occurring in other sectors are being extended into mineral exploration and mining. The novel approach taken by Rakai Resources is an example (Rakai Resources, 2014).

References

Ankli, R. E., & Varadan, S. (1999). Bre-X in hindsight: What we know about the Busang gold fraud. Retrieved July 31, 2014 from http://www.sbaer.uca.edu/research/sribr/1999/18.pdf.

Boutilier, R., & Thomson, I. (2011). Modeling and measuring the social license to operate: Fruits of a dialog between theory and practice. In *International Mine Management 2011*, Queensland, Australia.

Conciencia Minera. (2012). En el Día de la Minería, las comunidades originarias del Departamento los Andes, firmaron la "Declaración del Departamento los Andes". Retrieved June 28, 2012 from http://www.concienciaminera.com.ar/2012/05/en-el-dia-de-la-mineria-las-comunidades-originarias-de-la-puna-expresan-su-apoyo-a-la-actividad/.

de Echave, J., Diez, A., Huber, L., Ricard Lanatta, X., & Tanaka, M. (2009). *Minería y Conflicto Social*. Lima: Instituto de Estudios Peruanos (IEP).

Department of Justice. (2012). Canadian charter of rights and freedoms part I of the *Constitution act, 1982*. Retrieved June 28, 2012 from http://laws-lois.justice.gc.ca/eng/Const/.

Extractive Industries Transparency Initiative. The EITI principles and criteria. Retrieved December 28, 2010 from http://eiti.org/eiti/principles.

Frynas, J. G. (2005). The false developmental promise of CSR: Evidence from multinational oil companies. *International Affairs, 81*(23), 581–598.

Garvey, N., & Newell, P. (2005). Corporate accountability to the poor? Assessing the effectiveness of community-based strategies. *Development in Practice, 15,* 389–404.

Gobierno de Salta. (2012). Comunidades Collas de Los Andes y la Cámara de la Minería firmaron acuerdo de cooperación. Retrieved August 1, 2014 from http://www.salta.gov.ar/prensa/noticias/comunidades-collas-de-los-andes-y-la-camara-de-la-mineria-firmaron-acuerdo-de-cooperacion/17595.

Guay, L. (2012). Durabilidad, responsabilidad social y sociedades efectivas. Retrieved November 29, 2012 from http://www.olami.org.ar/archivos/eventos/DURABILIDAD,%20RESPONSABILIDAD%20SOCIAL%20Y%20SOCIEDADES%20EFECTIVAS.pdf.

Gunningham, N., & Rees, J. (1997). Industry self-regulation: An institutional perspective. *Law & Policy, 19*(4), 363–414.

Hutton, S., & Beaudry, P. (2014). Canada steps up the fight against foreign corruption. Retrieved July 31, 2014 from http://www.stikeman.com/2011/en/pdf/ALM500_ForeignCorruption.pdf.

IBA Research Network. (2010). Retrieved June 28, 2012 from http://www.impactandbenefit.com/.

Initiative for Responsible Mining Assurance. (2017). Retrieved March 6, 2018 from http://www.responsiblemining.net/.

International Development Research Centre. (2003). *Mining companies and local development— Latin America: Chile, Colombia and Peru. Executive Summary* (Mineral Policy Research Initiative report). Ottawa: International Development Research Centre.

International Finance Corporation (IFC). (2012). Performance standards and guidance notes—2012 edition. Retrieved June 28, 2012 from http://www1.ifc.org/wps/wcm/connect/Topics_Ext_Content/IFC_External_Corporate_Site/IFC+Sustainability/Sustainability+Framework/Sustainability+Framework+-+2012/Performance+Standards+and+Guidance+Notes+2012/.

International Labour Organization. (2008). Application of convention no. 169 by domestic and international courts in Latin America. Retrieved June 28, 2012 from http://www.ilo.org/wcmsp5/groups/public/---ed_norm/---normes/documents/publication/wcms_123946.pdf.

International Labour Organization. (2012). Convention 169. Retrieved June 28, 2012 from http://www.ilo.org/indigenous/Conventions/no169/lang–en/index.htm.

ISO. (2010). ISO 26000—Social responsibility. Retrieved February 6, 2015 from http://www.iso.org/iso/home/standards/iso26000.htm.

Kreuzer, O. P., & Etheridge, M. A. (2010). Risk and uncertainty in mineral exploration: Implications for valuing mineral exploration properties. *AIG News, 100,* 20–28.

Mas Herrera, M. J. (2005). *Desarrollo endógeno – cooperación y competencia*. Caracas: Editorial Panapo de Venezuela S.A.

Mining Association of Canada. (2010). Towards sustainable mining. Retrieved December 22, 2010 from http://www.mining.ca/www/Towards_Sustaining_Mining/index.php.

Mining Works for Canada. (2009). Strengthening our engagement with First Nations economies. Retrieved June 28, 2012 from http://www.miningworks.mining.ca/miningworks/media_lib/Newsletters/5311_MAC_MiningWorks_Vol_81_July1509_2.pdf.

Natural Resources Canada. (2015). Extractive Sector Transparency Measures Act (ESTMA). Retrieved January 5, 2016 from http://www.nrcan.gc.ca/acts-regulations/17727.

Newman, D. G. (2009). *The duty to consult: New relationships with aboriginal peoples*. Saskatoon: Purich Publishing Ltd.

OECD. (2011). 2011 update of the OECD guidelines for multinational enterprises. Retrieved June 28, 2012 from http://www.oecd.org/document/33/0,3746,en_2649_34889_44086753_1_1_1_1,00.html.

Ortiz Roca, H. (2012). Enfoque ético y aspectos humanos de la RSE. Retrieved November 16, 2012 from www.olami.org.ar.

Pontifical Council for Justice and Peace. (2005). *Compendium of the Social Doctrine of the Church*. Retrieved August 1, 2014 from http://www.vatican.va/roman_curia/pontifical_councils/justpeace/documents/rc_pc_justpeace_doc_20060526_compendio-dott-soc_en.html#Business%20and%20its%20goals.

Porter, M. E. (2011). Creating shared value: Redefining capitalism and the role of the corporation in society. Retrieved June 28, 2012 from http://www.isc.hbs.edu/pdf/2011-0609_FSG_Creating_Shared_Value.pdf.

Porter, M. E., & Kramer, M. R. (2011). Creating shared value. Retrieved June 28, 2012 from http://hbr.org/2011/01/the-big-idea-creating-shared-value/ar/pr.

Prospectors and Developers Association of Canada. (2008). Memorandum of understanding between the PDAC and the AFN signed by the AFN national chief and the PDAC president Toronto, March 4, 2008. Retrieved June 28, 2012 from http://www.pdac.ca/pdac/misc/pdf/080304-afn-pdac-mou-signed.pdf.

Prospectors and Developers Association of Canada. (2010). e3PLUS—A framework for responsible exploration. Retrieved December 28, 2010 from http://www.pdac.ca/e3plus/index.aspx.

Rakai Resources. (2014). Company overview. Retrieved August 1, 2014 from http://www.rakairesources.ca/about-rakai/.

Responsible Mining Index (2017). Retrieved June 3, 2017 from https://responsibleminingindex.org/.

Ruggie, J. (2008). *Promotion and protections of all human rights, civil, political, economical, social and cultural rights, including the right to development. Protect, respect, and remedy: A framework for business and human rights. Report of the Special Representative of the Secretary-General on the issue of human rights and transnational corporations and other business enterprises, John Ruggie* (No. A/HRC/8/5). New York: United Nations.

Ruggie, J. (2011). *Report of the Special Representative of the Secretary-General on the issue of human rights and transnational corporations and other business enterprises, John Ruggie— Guiding principles on business and human rights: Implementing the United Nations "Protect, respect and remedy" framework.* Retrieved June 28, 2012 from http://www.business-humanrights. org/media/documents/ruggie/ruggie-guiding-principles-21-mar-2011.pdf.

Sagebien, J., & Lindsay, N. (2011). Systemic causes, systemic solutions—CSR in a social and environmental value governance ecosystems context. In J. Sagebien & N. M. Lindsay (Eds.), *Governance ecosystems—CSR in the Latin American mining sector* (pp. 12–30). Houndmills, Basingstoke, Hampshire, U.K.: Palgrave Macmillan.

Salas Carreño, G. (2008). *Dinámica Social y Minería - Familias pastores de puna y la presencia del proyecto Antamina (1997–2002).* Lima: Instituto de Estudios Peruanos IEP.

United Nations. (2007). United Nations declaration of the rights of indigenous peoples. Retrieved June 28, 2012 from http://www.un.org/esa/socdev/unpfii/documents/DRIPS_en.pdf.

Vanclay, F., & Hanna, P. (2019). Conceptualizing company response to community protest: Principles to achieve a social licence to operate. *Land, 8,* 101.

Chapter 3
Theoretical Framework

The purpose of this study was to show how relationships affect the course of social events and the perceived present and future benefits and harms of mineral exploration projects. To arrive at a coherent, internally consistent analysis and interpretation of the results of the interviews, it is necessary to analyse them within an appropriate theoretical framework. This chapter describes first the place of relationships in the overall performance of a mineral exploration project, and second the theoretical framework that I selected.

3.1 The Performance of a Mineral Exploration Project

Given favourable geology and financing, three key factors of equal weight determine the overall performance of a mineral exploration project: knowledge/technology, governance and relationships, as shown in the diagram in Fig. 3.1.

While for simplicity of representation the diagram shows three separate parts, in reality each key factor permeates the others: governance cannot function without knowledge, technology and relationships; knowledge and technology require relationships and governance to produce results; and good governance, knowledge and technology promote relationship development. This book focuses on the relationship factor using the theoretical framework discussed below.

3.2 Theoretical Framework

As circumstances surrounding mineral exploration projects often fluctuate considerably over short time periods, I selected symbolic interactionism as a theoretical framework. It views the world as always "coming into being" and aims to explain the path that leads from individual interactions to decisions and outcomes. This approach

J. Boon, *Relationships and the Course of Social Events During Mineral Exploration*, SpringerBriefs in Geoethics, https://doi.org/10.1007/978-3-030-37926-1_3

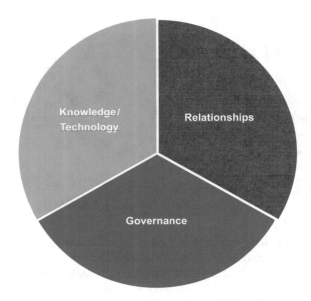

Fig. 3.1 Key factors influencing the performance of a mineral exploration project

is attractive in view of the varied and complex interactions between the diverse actors and groups of actors that are involved in mineral exploration projects.

Figure 3.2 shows a schematic of the symbolic interactionism sociological framework that I used to describe and analyse the social processes at work during a mineral exploration project (Blumer, 1969).

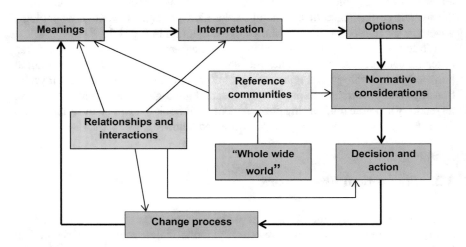

Fig. 3.2 Processes at work during a mineral exploration project

People give initial meanings (to the company; to mining; to the community) at the first encounter between the community and the company. They arrive at these meanings through interactions between fellow citizens on the community side and employees on the company side. Further, interactions lead to an interpretation of the situation, and to the development of options for action. They then weigh these options against the perceived norms of the actors' reference communities, make decisions and take associated actions. They can always redefine the situation and re-judge their action before taking it. Reference communities are the audiences of real or imaginary people whose conception or picture of communal life and associated values provide us with normative cues often in an almost subconscious way (after Athens, 2010). Examples of reference communities are the local general assembly (for members of Andean communities) and professional bodies or trade associations (for engineers and geologists). I used the plural because there is a reference community for each of the identities a person assumes.

Several factors, including continued interaction, can trigger change processes, and feed into a change of meanings and the cycle starts all over again. Relationships drive interactions, which in turn drive change, meanings and interpretations.

In the case studies, I arrived at the meanings the actors gave to things and people using contextual information (such as reports and articles) and interviewee comments. In various cases, the assumed meanings were reasonable assumptions based on the behaviour of those who assigned the meaning. As I mentioned earlier, the giving of meanings in many cases occurs semi-consciously, for which reason uncovering them from interviewee comments often involved a degree of interpretation. The concept "reference community" played a key role in the interpretation and analysis of the case study results.

Reflecting on the results of the present study it appeared that, when looking for a normative framework against which to judge their options for action, people often drew on more than one "reference community" of real or imaginary people. The relevant sections of a plethora of publications, databases, reports, newspaper articles, social media communications, workshops and television and radio programs (that also influenced the value system of the real people in the reference community) also influenced them. Therefore, the concept "reference community" as used in this book includes these additional sources. This broader reference community consists of subsets of particular communities of reference. These subsets are networks that are dynamic and ever changing, both in terms of their network properties and of their norms-related content. The decisions made by the persons drawing on the networks recursively modify their content. The more widely defined reference communities are a link between the micro-processes observed in the case studies and the macro-processes of society.

The wider reference community concept is particularly useful where a person can access a wide, sometimes global network. For example, in the Fruta del Norte case, the nodes in Aurelian/Kinross' network included the company's head office in Canada; community relations practitioners in sister companies in other countries and continents; the Mining Association of Canada; organizations such as the Global Compact, Organization for Economic Cooperation and Development and the Global

Reporting Initiative. These nodes were additional to the local nodes such as the parish council, cantonal, provincial and national governments and the Ecuador Council for Responsible Mining. Reference communities are dynamic and change with circumstances and time; members of different subsets (whichever way defined) within a larger group likely have different reference communities; and the less established and the more diverse the group, the more diverse and diffuse the reference communities related to the subsets within it. The weighing of decision options against the *perceived* norms of the reference community takes place "in one's head". As was mentioned earlier, reference communities consist of real and imaginary people and it should be possible to trace the networks of the real people. An analysis of incomplete networks can provide useful insights, and informal network tracing approaches avoid the ethical challenges that the use of a detailed social network analysis would pose (Borgatti & Molina, 2003).

3.2.1 Transactional Needs

The "transactional needs" individuals have in all encounters with others strongly influence interaction. The most important of these needs relate to confirming one's identity.

Each person has various identities. In order of decreasing importance, these are:

- core identity: the image and emotions that people have about themselves and that they carry into most encounters.
- social identity: self-conception related to belonging to categorical units such as race, gender and culture.
- group identity: self-conception related to membership in groups, organizations and communities.
- role identity: self-conception related to the roles individuals are assuming such as teacher, father, wife or professor.

The emotional intensity experienced in the encounter and the range of situations to which it applies decreases from the beginning of this listing, whereas the level of conscious awareness of the participants of their need increases from the beginning. Additional transactional needs include (Turner, 2011):

- benefits are drawn from an encounter
- inclusion
- trustworthiness of the other party
- transparency.

Meeting these needs is at the core of establishing relationships and all actors should consider these. Four additional tactical dimensions add a useful practical component (Gawley, 2007; Henslin, 1968):

- being visible, maintaining a presence and establishing routines

- expressing sincerity
- personalizing encounters and
- clarity about your own trust expectations.

Maintaining a presence and establishing routines are key factors in this respect. As will be seen later, these routines can lead to the formation of new social structures. The degree to which the parties meet transactional needs is a crucial factor determining the characteristics of the relationships that result from the interactions of the parties, as expressed by the relationship indicators that I will discuss later in this chapter.

It is possible to extend the above model for interactions between individuals, to interactions between groups. Blumer argued "…group action takes the form of a fitting together of individual actions. Each individual aligns his action to the actions of others by ascertaining what they are doing or what they intend to do—that is by getting a measure of their acts (Blumer, 1969)". Large-scale group identity can be linked to an interpersonal network of members who share a category of membership (linked to a certain concept, field, issue, political direction…) and who provide support for the group identity (Deaux & Martin, 2003). For example, the loosely coordinated group that was opposing mining in the province of Azuay in Ecuador had adopted as its identity the slogan "El agua es la vida" (water is life). Mainly its leader promoted the role identities that supported the formation of this identity, and the group drew on an interpersonal network of others that acted as its reference community. Through a fortunate coincidence, a series of photographs of speakers that took part in the "referendum" on mining that this group organized provided a graphic representation of a number of the nodes in this interpersonal network: members and leaders of a number of indigenous organizations, some politicians and others (Flickr, 2011). The interplay between the identities of group members can cause change that can determine the course of social events.

The processes described above are dynamic and recursive and work through relationships, and meanings, identities and reference communities continually change. The relationships, interactions and interpretations at work in mineral exploration situations are the dynamo that drives the unfolding of social events. While symbolic interactionism focuses on "micro" interactions between individuals, these manifold interactions combine to create social structures that themselves affect meanings. Individuals are simultaneously creating social structures and that at the same time change them (see also Nugus, 2008). This applies to all actors involved: the actions of company personnel, authorities, community members and the social structures created can have either positive or negative effects. For example, the case studies that I will discuss in Chap. 4 showed that in the Loma Larga project and the Fruta del Norte projects they strengthened the communities of San Gerardo, Chumblín and Los Encuentros. In the Río Blanco case that I described in my Ph.D. thesis, the new social structures that formed led to spiralling polarization and violence (Boon, 2015).

3.2.2 Relationships

A relationship exists if actions of one of the parties involved affect the behaviour and actions of one or more of the other parties. Relationships drive the social interactions that lead to meanings. They form in the minds of the parties and provide the framework in which their interactions take place, and they evolve over time. Relationships can be positive or negative. Relationships, as a social construct, are always embedded in a context (Atkinson, 2004). History, prior information and reputation of the parties play an important part, and Ferris et al. described the stages of development of a relation (Ferris et al., 2009). This book views the relationships in the context of mineral exploration and the communities, authorities and companies potentially affected by it, and of the related factors and circumstances. Aspects of relationships mentioned in the PDAC e3 Plus tool include trust; respect; communication; mutual benefit; agreement on the roles of the partners; and conflict resolution (Prospectors and Developers Association of Canada, 2010). Whilnot specifically using the word "power", the guidelines suggest mechanisms to provide resources to communities that allow them to negotiate on an equal footing.

I developed a tool to characterize relationships. The tool assigns a qualitative "risk-of-conflict measure" to a series of relationship indicators, which allows identification of areas that need attention. I derived a number of these indicators from the e3 Plus guidelines. I added mutual understanding as an additional "check" on the quality of communication; and I added goal compatibility, frequency, stability and productivity as these could affect the course of social events. While each indicator focuses on a different aspect of a relationship, the indicators are interdependent to a certain extent. Table 3.1 shows the resulting list of indicators together with a brief description of each.

Below I illustrate the methodology I used for characterizing the bilateral relationships for the cases that I will describe in Chap. 4 of this book. I based detailed descriptions of bilateral relations on meticulous reading of interviewee comments and I shared the detailed case study reports with the respective projects to check for factual errors and to meet the conditions under which the company provided access to its personnel and premises. The descriptions of the bilateral relationships formed the basis of a "relationship matrix" the cells of which represent bilateral relationships.

Table 3.2 shows Kinross' Fruta del Norte project in Ecuador as an example. The hues of red and green for each cell show its degree of "negativity" or "positivity", respectively. The community interviewee comments shown below formed the basis for the green hue of the cell representing the bilateral relationship between Kinross and "general community" (row 13, column 4, Table 3.2). For brevity and ease of reading, I selected representative comments here and in other citations. My thesis provides the complete comments (Boon, 2015):

> …Now people are in favor of the company although there are persons who are against mining…The moral thing is very important! Kinross managers are people with values…Kinross had a big impact on the community, but actually there is very little opposition to its work…The relation between the president of the parish council and the management of Kinross is very much based on respect and trust, which makes that community objectives are

achieved…The community will not allow a company other than Kinross to enter … Getting to know how Kinross is managed helped grow support for mining and for the company … The different mentality of the people working in Kinross contributed to better management with the community… The key values of the company are that people come first, and it considered differences to be strength… The company treats its workers fairly – it would prefer losing money over losing people … They care very much about the people of the region and take a wide range of aspects into account [male community members of various walks of life]…

While the majority of community comments were positive, there were a few complaints as well:

…I believe that they exclude the Shuar [indigenous] community from employment…The connection should be like it was with Aurelian, there is no communication now [female Shuar community member]…

The red hue assigned to the cell representing the relationship between the governor of the province (representative of the national government) and the provincial prefecture (allied with a different national political party) was based on press and Internet reports and on an emphatic comment of an interviewee closely linked to one of these offices:

…The relationships between the Prefecture and the Governor's office are bad…

Table 3.1 Indicators for characterizing relationship

Indicator	Description
Trust	To believe despite uncertainty. It involves taking risk and beliefs about expected behaviours of the other
Respect	To take notice of; to regard as worthy of special consideration
Communication	Hear and being heard
Mutual understanding	Degree to which each side can correctly express what the other side is saying
Conflict resolution	Degree to which conflict resolution mechanisms exist and are productively used
Goal compatibility	Degree of compatibility between the goals of the parties (to what degree achieving the goals of one party supports or hinders the achievement of the goals of the other party). It is possible to achieve two groups of completely different goals through the same project (e.g. the construction of a dam to generate electricity at the same time creates a lake that offers opportunities for fishing and tourism)
Balance of power	The extent to which each party influences the other and to which each party affects final outcomes
Focus	Clarity about who should be legitimately involved in the relationship and about the matters at stake in the relationship
Frequency	Frequency of significant interactions between the parties (whether positive or negative)
Stability	Degree of predictability
Productivity	Degree of achievement of target results

Table 3.2 Relationship matrix for the Fruta del Norte case

1	Provincial prefecture	8	Water Council	15	Artisanal Miners Association	22	Artisanal Miners Association
2	Mayor's office	9	Shuar Federation	16	Pick-up truck cooperative	23	Educational institutions
3	Parish Council	10	Catholic Church	17	Women's organization		
4	"Community" (general sense)	11	Political lieutenant	18	APEOSAE		
5	Communities of Los Encuentros	12	Governor	19	Cattle Association		
6	National government	13	Kinross	20	Volunteer Group		
7	Political parties	14	Employee Association	21	Sports clubs		

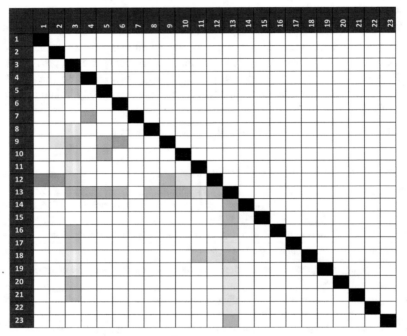

Green represents a positive relationship and red represents a negative relationship. The intensity of the colour indicates the perceived intensity of the relationship. For the blank cells, either a relationship did not exist, or no relevant information was available

I used a similar process to assign hues to the other cells shown in the example table. The matrix clearly shows that Kinross and the parish council were the main relationship axes, followed by the Shuar Federation. I produced such a matrix for each case study that provided an immediate impression of the most important actors and their relationships. The relationship matrices allowed me to see at a glance both the number of actor groups involved and the actor groups that had the highest number of connections to other actor groups ("relationship axes"). In almost all cases, the company and the local authorities constituted the main relationship axes.

As the relationship axes did not differ greatly between cases, I have not included the matrices for the individual cases in this book.

Contextual factors can significantly influence relationships and interactions. The physical environment can have a strong influence, for example, the steep slopes of the Andes make for a huge vertical variation in climatic conditions that links altitude to the plants that can grow there, the animals that survive there and the types of activities in which people are engaged. It also affects the frequency of interactions: people living in the same altitude zone can interact daily, whereas contact between people living at different altitudes will be the less frequent the greater the separation in altitude. Other physical factors that influence daily life and therefore relationships, interactions and meanings include rivers, lakes, mountain ridges, slope orientation, annual precipitation rate and soil types. Prior history can have a determinant effect, for example, the images of colonial mining in Peru and of the environmental damage wrought by the Cerro de Pasco Corporation in La Oroya in the early parts of the previous century are part of the collective psyche of the people of Peru and inevitably affect the meanings they give to mining. These images are also part of a reference community in the sense that they constitute a "negative" norm.

Availability and spatial distribution of resources also can play a role. For example, a significant part of relationship patterns in some case studies concerned management of irrigation and potable water supplies. These patterns in turn affected interactions and meanings, which in one parish led to a decision to take forceful action against the mineral exploration project and those who supported it. Physical proximity promotes interaction and meaning formation: the greater the physical distance, the lower is the likelihood of interaction and meaning formation. In addition to physical proximity, social distance can play a part. Social distance can be thought of as the property of an object "other", for example, of an out-group, about which people tend to think in a more abstract manner. It can also be a consequence of a person's mindset: she or he may think about the world in a more distant manner (Lammers, Galinsky, Gordijn, & Otten, 2012). The farther socially away a person or entity, the less likely it is there will be interaction and meaning formation through interaction.

These contextual influences can lead to the formation of "subunits" within which meanings, interpretations and decisions converge around a "majority position" that is more amenable to tracking than are individual meanings, interpretations and decisions. Subunits can be organizations, cultural groups, neighbourhoods or entire communities, depending on the case. Subunit members whose meanings, interpretations and decisions have not converged around the majority position can become catalysts of the process of changing meanings.

References

Athens, L. (2010). Naturalistic inquiry in theory and practice. *Journal of Contemporary Ethnography, 39*(1), 87–125.

Atkinson, S. (2004). Senior management relationships and trust: An exploratory study. *Journal of Managerial Psychology, 19*(6), 571–587.

Blumer, H. (1969). *Symbolic interactionism*. Englewood Cliffs, NJ: Prentice-Hall Inc.

Boon, J. (2015). *Corporate social responsibility, relationships and the course of events in mineral exploration—An exploratory study* (Unpublished Ph.D. thesis). Carleton University, Ottawa. https://curve.carleton.ca/system/files/etd/6c6598d4-c436-409e-9ba1-40dea2d37d2c/etd_pdf/7b39ca613ff7e7e2df52ed82580e3974/boon-corporatesocialresponsibilityrelationships.pdf.

Borgatti, S., & Molina, J. (2003). Ethical and strategic issues in organizational social network analysis. *The Journal of Applied Behavioral Science, 39*(3), 337–349.

Deaux, K., & Martin, D. (2003). Interpersonal networks and social categories: Specifying levels of context in identity processes. *Social Psychology Quarterly, 66*(2), 101–117.

Ferris, G., Liden, R., Munyon, T., Summers, J., Basik, K., & Buckley, M. (2009). Relationships at work: Toward a multidimensional conceptualization of dyadic work relationships. *Journal of Management, 35*(6), 1379–1403.

Flickr. (2011). Sistema comunitario agua potable VP-T. Retrieved October 21, 2013 from http://www.flickr.com/photos/62081634@N05/6215588294/in/photostream/.

Gawley, T. (2007). Revisiting trust in symbolic interaction: Presentations of trust development in university administration. *Qualitative Sociology Review, 3*(2), 46–63.

Henslin, J. M. (1968). Trust and the cab driver. In M. Truzzi (Ed.), *Sociology and everyday life* (pp. 138–157). Englewood Cliffs, NJ: Prentice-Hall.

Lammers, J., Galinsky, A., Gordijn, E., & Otten, S. (2012). Power increases social distance. *Social Psychological and Personality Science, 3*(3), 282–290.

Nugus, P. (2008). The interactionist self and grounded research: Reflexivity in a study of emergency department clinicians. *Qualitative Sociology Review, 4*(1), 189–204.

Prospectors and Developers Association of Canada. (2010). e3PLUS—A framework for responsible exploration. Retrieved December 28, 2010 from http://www.pdac.ca/e3plus/index.aspx.

Turner, J. H. (2011). Extending the symbolic interactionist theory of interaction processes: A conceptual outline. *Symbolic Interaction, 34*(3), 330–339.

Chapter 4
Case Studies

For this book, I selected five of the nine case studies I conducted in the hope that they will provide useful touchstones for the reader. My Ph.D. thesis describes the details of all nine case studies (Boon, 2015). For each case, I present the following information: field study details; project description; context; patterns and characteristics of relationships; company social responsibility approach; and perceived present and future benefits and harms at the time of the study. I finish each case study with a brief comment on the status in September 2019 of the project.

The socio-economic indicator, "unmet basic needs", occurs in the context section of many of the case studies, and its definition as provided by the Instituto Nacional de Estadística y Censos is (INDEC, 2001a):

> Basic needs are unmet if at least one of the following conditions applies: (i) 3 or more persons per room; (ii) homes in an "inconvenient" type of housing (rented room; precarious or other type of housing, excluding a house, apartment, or hut); (iii) homes without any type of toilet (iv) presence of children between 6 and 12 years of age not attending school; (v) households with four or more members per working person, and where the head of the family has less than grade three education. The index of unmet needs that is derived from these variables ranges from 0 (no unmet needs) to 1 (all needs are unmet).

4.1 Research Method

The Cauchari-Olaroz field study was part of an evaluation of "e3 Plus—A Framework for Responsible Exploration", the CSR tool that the Prospectors and Developers Association of Canada developed (Prospectors and Developers Association of Canada, 2010). The PDAC prepared an interview guide that paralleled e3 Plus. For the Lindero case, I developed a new interview guide more closely aligned with the relationship focus of the present study. With the assistance of staff from the Instituto Nacional de Investigación Geológico, Minero y Metalúrgico del Ecuador

(INIGEMM), I expanded it for use in the other case studies. I added a specific question about reference communities to the interview guide for the Palma case, as the importance of the reference community concept had become clear at that stage.

Interviews followed a semi-structured format, with participants having much leeway to expand on topics of interest to them. I recorded respondents' comments in writing during and immediately after the interview and later entered them into MS Word documents. I coded these documents using Atlas.ti software and used the resulting code structure and outputs to build a detailed report for each case study (Atlas.ti, 2012). Even though there were differences between interview guides, the comments provided by interviewees were sufficiently wide-ranging to allow coding using the framework implicit in the structure of the case study reports.

4.2 Palma

4.2.1 Field Study Information

Interviews took place between 6 July and 10 July 2013 (inclusive), and 30 people took part, six of whom were Empresa Chungar personnel and 24 were members of the comunidades campesinas Santa Rosa de Chontay, Cochahuayco, Villa Pampilla, Sisicaya and Santo Espíritu and of the Centro Poblado Palma, Chilllaco, Antapucro and Nieve Nieve. As mentioned earlier, I added a question on reference communities to the interview guide for this case study. Interviewees included both proponents and opponents of mining. I conducted all interviews, with financial support from the Organization of American States.

4.2.2 Project Description

Mining activity in Palma dates from before 1985. In 2009, Volcan Compañía Minera S.A.A (a Peruvian company), through Empresa Administradora Chungar S.A.C., acquired the concessions and started a drilling programme in the third quarter of 2011. The Palma project was studying a lead–zinc–barite deposit (Volcan Compañía Minera S.A.A, 2010). It is located on the left bank of the Lurín River, in the district of Antioquía, province of Huarochirí, Lima Region, at about 60 km east of Lima. The Palma project was small. Its governance consisted of the project manager and two managers who reported directly to him. The managers oversaw the work of consultants. In this sense, it was an "administrative" project. Many of the employees of the consulting companies were from the area, and they identified with the work in which they had specialized such as sampling, photography, and classification of drill cores. Daily meetings in which the exploration team members discussed safety and security, tasks, activities for the day and evaluation of the previous day were an important part of project governance, facilitating communication and cooperation.

The settlements closest to the project are the Centro Poblado (C.P.) Palma at a distance of four km, and Chillaco at 7 km. The land belongs to the Comunidad Campesina (C.C.) Cochahuayco (Volcan Compañía Minera S.A.A, 2013). Comunidades campesinas are "…organizations of public interest, with legal status and juridical personality, consisting of families that live in and control certain territories. They are linked through ancestral, social, economic and cultural bonds, expressed through communal property of the land, communal work, mutual assistance, democratic government and development of multi-sector activities…Annexes of the Community are the permanent settlements located in communal territory and recognized by the General Assembly…" (Centro Peruano de Estudios Sociales (CEPES), 2010). A Centro Poblado is "…any place in the national rural or urban territory, identified by a name and permanently inhabited" (Gerencia Regional de Planeamiento, 2006). In the study area, most Centro Poblado form part of a Comunidad Campesina, with the exception of Palma, that occupies land belonging to three comunidades campesinas: Cochahuayco, Sisicaya and Espíritu Santo.

4.2.3 Context

The local economy is land-based, and the farms in the valley produce apples, quince, avocadoes and vegetables sold at local markets and in Lima. In 2009, 15.6% of the population was living below the poverty line that was located at S/. 250 per month (approximately C$100) (Quispe López, 2009). The district municipality is the most important local governance entity, as it is the channel for infrastructure project budgets. A community interviewee mentioned that in the past, some mayors used the budgetary processes to buy votes.

> …The previous mayor bought votes (by giving cement and other materials to people) and people became used to it. It was a waste of resources. The present mayor does not do this [female community member]…

The comunidades campesinas are second in importance—they manage lands and social issues. Their presidents are the most important persons in the communities, although the general assembly that has supreme power prescribes their actions. As is to be expected, there were differences in quality of governance between communities. Both presidents and town councils are subject to criticisms, the most common one related to not carrying out their duties and the most serious one to accepting bribes:

> … There is envy between the presidents of the communities in the valley … The President's meetings do not have an agenda and he is disorganized – much time is wasted [female community member]… The community is not learning – the president is bad. We have not improved at all in two years …The president is on balance so-so. He relates with the company…The president looks out for the village in dealings with the mining companies…The president is not very interested … The assemblies function well – the majority of people

are interested and the decisions are accepted …The community dismissed the town council because they were implicated: the company gave $2000 to each of them not to denounce the company for exploitation without a permit [male community members]…

The administrative structures under which the comunidades campesinas operate play a key role in the functioning of local society and the general assemblies discuss any matter. Their minutes are legal documents that record decisions and spell out who should take what actions. The existence of two classes of citizens, pobladores and comuneros, is an interesting aspect of governance that dates back to Inca times. It has been "re-officialized" since 1920 (Del Castillo, 2006), and the associated legislation is regularly updated. Comuneros are those born in the comunidad campesina, the children of comuneros and people who become members of the comunidad campesina through a formal process. Only qualified comuneros have the right to vote or can run for election. "Pobladores" are people living in a community who are not comuneros. The general assemblies and the associated town councils of the comunidades campesinas have no jurisdiction over the pobladores while the district municipality has jurisdiction over both pobladores and comuneros.

Many citizens contribute directly to the functioning of the communities through their participation in communal tasks, for example construction of irrigation canals, cleaning up the town in preparation for the patron saint celebrations, and the construction of speed bumps. A committee established for the purpose manages each task. There was neither police nor a fire brigade in the district, but the governance mechanisms described above appeared to cover these functions at least partially. Land is communal property, and the comunidad campesina assigns lots to the comuneros on which to build their houses. The comunidad campesina manages the use of the communal lands, which for the majority of communities are officially registered. The lands of Cochahuayco and Villa Pampilla had not yet been registered pending resolution of ongoing litigation. The exploration camp was located on rented land in the Cochahuayco part of C.P. Palma. There were also private lands, for example the C.P. Antapucro consists of pobladores who own private property within the C.C. Sisicaya, and different rules apply.

Each community manages its own potable and irrigation water supplies through corresponding water boards. Communities seek support from the municipality, and some have received support from international organizations (e.g. in Cochahuayco, the Japanese aid agency has supported the construction of the potable water system and a Spanish NGO has helped with construction of the jam and juice plant). In many communities, local action has been instrumental in obtaining electricity and other infrastructure such as Bailey bridges and extension of school buildings.

4.2.4 Observed Patterns and Characteristics of Relationships

The Palma project operates in the context of a scandal surrounding the Huascarán project, a previous Volcan exploration project in the Lurín valley close to C.C. Espíritu

Santo. The regulatory agency shut down Minera Huascarán's project because of environmental and permitting violations, and the aftermath of this corruption scandal did not receive satisfactory attention. There were also accusations of Minera Huascarán having fraudulently obtained town council member signatures on official documents.

As mentioned earlier, general assemblies are an important governance tool. The following interviewee comments illustrate aspects of the way they operate:

> …In the assembly I defend education – I bring issues forward [male member of the parents association] …. Close to 60–70 people attend – for a total population of more than 100 this is less than half…There are 5–6 assemblies per year [other communities reported higher participation rates]. Most often, they take two to four hours. Usually they arrive at reasonable agreements [female community members]…The assembly has supreme power, the president depends on the assembly…Through the assembly the presidents of the associations act as such, or they act as individual members, depending on the issue…All relationships are important – we are in contact every day …Conflicts are resolved through the assembly – the "fiscal" is charged with that [the "fiscal" is the member of town council responsible for keeping order and apply the rules]. There are statutes and regulations that include fines, penalties, calls to attention, and communal work. The members of the executive are juridical persons. This functions well … I am a member of the mining committee of the assembly…There are extraordinary assemblies, for example when we decided to form our own council [not all Centro Poblado have a full council]. Our requests for support to the company are submitted in writing and are part of the official minutes of the assembly… There was a "careo" [face-off between witnesses] in the general assembly [related to the Huascarán corruption case]. Neither of the two sides could prove their version of events…We had an extraordinary assembly 4–6 months ago to revoke the permit for the Road of the Inca to the National Culture Institute. … Conflicts about land between comuneros are resolved by tribunals of the assembly. The general law of comunidades campesinas regulates all this There is a general assembly each month and usually more than 40 comuneros take part. Quorum for the general assembly is one half plus one [male community members]…

For conflicts that cannot be resolved through the mechanisms available to the general assembly or through conversations, the lieutenant governor and afterwards the judge of the peace (who mediates in any type of conflict) come in. If the judge of the peace cannot resolve a problem, he or she hands it over to the "judicial chain" that extends to Lima. Land issues are a frequent source of conflict, often caused by lack of registered title:

> …There always has been much conflict between the comuneros of Sisicaya, almost always related to land. As they do not have title the issues are very difficult to resolve and there are always people who take advantage of others [male company personnel]…

Many conflicts do not reach the general assembly and rarely get resolved—they remain hidden and latent. One community member opined that it is important to be open about problems and discuss them; otherwise, there will not be a solution:

> …Not everything functions perfectly – there will always be inconveniences. For example, some families say nothing [even when there is a problem] and nothing gets resolved. However, there are people who talk [male community member]…

There is friction between the different levels of authority of the municipal district:

> …There are always differences of opinion between the municipality and the Comunidad Campesina which leads to conflicts [male community member]…

The relationship axes of the bilateral relationship matrix involved the judge of the peace and the governance structures of the comunidad campesina, with relationships linked to Empresa Chungar representing a secondary axis. As mentioned earlier, there was a certain degree of friction and envy between all administrative units, be they comunidades campesinas or centro poblado. In most communities, everyone knows everyone and several interviewees commented that all relationships, while different, are important:

> …All relationships are important, but they are different. The villages prioritize as a function of customs, culture, feelings, not in function of necessity [male company employee]…All relationships are important…All relationships are very important…In the village the day-to-day relationships are friendly…The entire village is like a family – family relationships are also important [female community members]….

There were many connections, and there was much mutual support and cooperation (albeit some of it under pressure):

> …Everything is interlinked and there is also support between community members in case of illnesses etcetera. There is no problem with working together: if someone does not work, he is fined. If he doesn't pay his fines three times he is excluded from the community until he pays up [this refers to "faenas": chores imposed by the committee that organizes them] [male community member]…

With respect to the relationships between the company and the community, the contract employees who are comuneros of Cochahuayco played an important role:

> …We are ambassadors and information channels in two directions [male company contractors]…

The patron saint festivities played a very important role in community life into which the community invested much and in which everyone participated. There were elaborate structures in place to support the preparations and the festivities themselves, and every community member has to do his or her share, as the following interviewee comments indicate:

> …The 'cajueleros are the people who contribute most to the communal festivities and they are part of the community president's function. In the procession they carry images of the saints … They belong to the Association of Cajueleros and are devout persons who take on responsibility for certain aspects of the festivities that often coincide with the role of mayordomo…The mayordomo takes care of the expenses of the patron saint's festivities, in exchange of a benefit (for example a lot for a house - some 100–200 m^2) or for farming (a hectare)…Everything is free: drinks, communal meal etcetera. For days for which there is no mayordomo, an S/. 100 fee is levied on everyone…Both the company and the community contribute to the festivities: drinks, food and music…The community organizes it – the comuneros living on communal lands are obliged to cook the meals…All lands are communal [male community members]…

Several interviewees identified two types of comuneros: those who live in Lima and those who live in the community. The former appeared to be causing problems:

> …There are two groups of comuneros: those that live in Lima and those that live here. The first group is always looking for money, but the company does not give money, it gives social

and other support. The first group tries to convince the second group that they should provide them with benefits. Because they are more erudite, they are agitators. This is a problem – we need to make it clear to them that we are here for the communities… We have a major challenge with the information and manipulation that comes from outside, for example by the children of comuneros that live in Lima [male company personnel]…

Comments about the general characteristics of people in the area showed some consistency, and it is worth summarizing these as they provide a context for the relationships and interactions that drive the symbolic interactionist processes. The positive comments say that the people are humble and unassuming, and that they are happy to live without money:

…They live from what the earth gives them and have simple goals…They receive you with open arms…They are cooperative [male community members]…

These same characteristics also have negative aspects from the point of view of community development:

…They have no initiative, they have a fearful attitude and it is as though they were living in a cage. There is lack of interest, lack of trust, and fear. They could organize themselves, but half of the people are slackers and they like being drunk…In addition, there is much envy between Centro Poblado, annexes and Comunidades Campesinas. They distrust people from outside, because of jealousy. They are not building their capacity, they do not seek, they do not go out and are conformist …They do not want to do things right…They are always dissatisfied – would this be hereditary?…For these reasons many things cannot be achieved…Self-respect is very low which leads to lack of trust and to the formation of small groups led by those who have a little more knowledge [male community members]…

These interviewee comments agree with the observations of Claverías Huerse and Alfaro Moreno on the problems the agricultural producers in the Lurín watershed have with creating spontaneous productive chains. "…They do not manage to build networks or producer organizations to work together and identify strategic objectives, and they do not associate because they do not trust each other or other agents in the chain. In addition, they do not manage to orient themselves towards the new demands of the markets. The producers do not have the confidence to associate, there is much distrust between them and there are divisions in the community. The parents are very traditionalist, there is no attitude of generalized change, and the young are the only hope. The young who want to improve their situation are educating themselves for change, but the parents do not worry about change, they are very conformist" (Claverías Huerse & Alfaro Moreno, 2010).

4.2.5 Company Social Responsibility Approach

The official policy of Volcan Compañía Minera has the following areas of focus (Volcan Compañía Minera S.A.A, 2012): development and strengthening of sustainable productive capacities, support to education and health, promotion of local employment, support for basic road and public services infrastructure, institutional support, and promotion of culture. Volcan Compañía Minera only began paying attention to social responsibility half a decade before the present study, and the meaning given to social responsibility by many within the company was that of a cost item rather than an investment opportunity. For example, the reference community of the people in the finance department was that of financial accountants' professional organizations whose worldviews do not (yet) include social aspects of the business. The community relations officer was of the opinion that changing the company mindset was a greater challenge than establishing productive relations with the communities. In comparison to other case studies, the community relations aspect of the Palma project did not have many resources at its disposal.

The principal social objective of the Palma project was:

> …Not to change the people's customs, to maintain friendly relations and to develop productive products with the town councils in health, education, and economy as a first stage [male company personnel]….

The community relations officer conducted regular analyses of Strengths, Weaknesses, Opportunities and Threats on which he based the focus of his efforts. The Palma project designed its strategy using his advice and prepared an activity chronogram that was coordinated with exploration management in Lima. Their social responsibility programme was at an early stage:

> …We have not yet done much in infrastructure. Up to now we have made modest contributions but we have given social support and have been helping the communities to organize for example obtaining land titles in Cochahuayco and Sisicaya through technical (cartography contracts) and legal assistance…The project's contributions so far include: education: improve logistics, supply materials, coordination with educational units in financial support for capacity building, and paying more than half the salary of the teacher of the PRONOEI in Centro Poblado Palma [PRONOEI Programa No Escolarizado de Educación Inicial – a program aimed at children that have no access to formal education. Its philosophy involves assigning an important role to the community – (Unidad de Desarrollo Social)]; a health centre; social life: support for mother's day, Christmas festivities, seniors, relief in case of flooding; development (for example cement for irrigation canals; courses in raising animals and environment; the chapel; lighting in a hall); assistance with registration of land titles; sporting events: organized in cooperation with all communities to avoid intercommunity politics from entering into play [male company personnel]…

4.2.6   Perceived Present and Future Benefits and Harms

The communities had carried out various development projects without company support, for example a preserves processing plant, electrification, a bridge and improvements to the school building

> …We constructed the plant with the help of the NGO Centro de Investigación en Educación y Desarrollo (Centre for Research and Development) and the Spanish NGO CODESPA that obtained funding from the city of Valencia in Spain… The plant belongs to the community…We have managed to obtain light [electricity], water and the Bailey bridge – through our

own effort…Now we have a school building with two floors and we can do more…We constructed a lookout with a dining hall, a kitchen and a stage, with money from the community. It was a large construction and many say that it cost almost 50% more than was budgeted. Budget control was bad [male community members]…

Various communities had additional development ideas:

…The most important issue is improvements to the irrigation systems – we want to extend the canal by 5 km… The assembly presented a request for support to the municipality, signed by each comunero. They are giving the support bit by bit and it could take two to three years…We need a hotel and sanitary services for tourism…need an antenna. The landline telephone works well [male community members]…

The Palma project manager was considering contributing technical assistance to the design of more sophisticated irrigation systems. There was a need for capacity building, but barriers included lack of interest, lack of confidence, fear and lack of credibility of some of the institutions that offer it:

…The community is building capacity. We want it to grow. The mining company is involved in capacity building. More resolve and action is needed. Growth is necessary. Knowledge improves things – theory, practice and deeds…There is no capacity building in the Centro Poblado… The town council members are not building their capacity. An NGO offered a leadership course but very few attended…There is lack of interest, lack of confidence, fear…The Ministry of Agriculture offered courses in health service, agriculture and animal husbandry, but people did not participate because the Ministry has low credibility [male community members]…

The observations made by Claverías Huerse and Alfaro Moreno that I mentioned earlier played a role (Claverías Huerse & Alfaro Moreno, 2010). The Palma project did offer regular training sessions to its employees and contractors, and some capacity building in the communities.

There were divided opinions on the benefits derived from the project. Some thought that the project had brought many benefits, some that it had brought some benefits and some that it had brought no benefits at all. Much of this variance likely relates to the degree to which the CSR activities of the project touched the interviewees or their associates or to their employment by the project. Those who felt there had been benefits pointed to contributions to the local economy, education, infrastructure and assistance with sorting out land titles:

…Many have benefited…Yes it has brought benefits…Sales stands, lodging, work, -very good…Benefits: the teacher's salary; they provide their staff with accommodation in the Centro Poblado and with food through the dining hall run by locals…Yes, they have helped with several things…They repaired the road…The company is advising us on land titles in Cochahuayco and Sisicaya. Once we have titles there will be much less conflict [male community members]…

However, others were not so sure:

…Very little benefit. Small things compared to the margin of capital they have. They are not tackling problems with great reach…They are beginning – there has not been much benefit. Many ask for reflection…There are advantages (support) and disadvantages. I am not sure…I do not know. The community relations officer is supporting us. There have not been disadvantages [male community members]…

…The arrival of the company did not have much effect – we are six villages away from the project – it is far…No damage so far – only minor incidents…Until now there have not been benefits. In general no. There are speculations that a certain group has benefited, but this could not be true…So far we have not obtained any benefits, but everything is going well [community members]…

Others were looking forward to a mine opening for the employment opportunities it would provide, and members of communities in the indirect zone of influence complained of lack of attention.

…Once there is a mine there will be funds for the mayor and there will be employment [female community member]…A responsible mine will be a source of employment [male community member]…

Several interviewees perceived that the company had benefited:

…The rented land and the exploration permit – through the agreement…Three years without a single problem, positive reputation of being a responsible company…They haven't had conflicts because of the work of the community relations officer…The signed agreements allow it to work with more confidence [male community members]…Community Relations is negotiating continually and the agreements assure tranquility. They have created the environment for being able to do the work…Too many requests could pose a risk but it is not a risk now [male company employee]…

The major concern of the majority of interviewees was the risk of contamination of the water and possible negative effects of a mine on agriculture:

…Worries remain – contamination needs to be avoided, but there has been none so far…Agriculture has to be 100% protected…Mining has lacked a conscience – it can contribute to agriculture…Often mining companies destroyed everything and left. We have to avoid that now…The Incas mined gold and did not cause contamination…I visited a mine and saw the containment dikes. I am afraid of the effects huaycos could cause [a huayco is a mud slide caused by rain – in the Lurín valley there are huaycos every year and about once every ten years there is a very severe one that could close off valleys. The name of the town of Cochahuayco is Quechua for "mudslide lake"] [male community members]… The tailings will poison the river – we depend on the water…There is a risk of contamination and we will not have prawns in the river any more [multiple additional comments related to concerns about water purity]… To this day I have the feeling that they are fooling us…We fear that we will be cheated. We will take everything with caution [female community members]…

Some interviewees saw independent monitoring of water and air quality as a way of dealing with the risk of contamination. They also realized that their own practices were contaminating the river:

…We are planning independent monitoring of water and air quality. We need support from the municipality, the Ministry of Energy and Mines and the Ministry of the Environment… We have to defend nature and the life that it gives us. We ourselves are contaminating [male community members]…

Several interviewees thought that they did not receive sufficient information about the Palma project's operations:

…We are close to the plant – we need information about what is going on there. Often we know nothing, even though we are close…If there is no information, we are not going to take the risk…I know that Empresa Chungar exists, but we have no information. There could have been much more [female community members]…

Interestingly, several interviewees volunteered suggestions that in their opinion would benefit the company. Most linked in one way or another to communication and involvement. While the Huascarán scandal continued to weigh on their minds, several of them suggested approaches that would help clear the air.

…There has to be a benefit for all, and that there be regular contact with *all*…Hold meetings whenever possible…Bring in contractors who are sensitive and not authoritarian…Many things need to be done and it would be good if the company became involved…I am sure that, if the new manager [i.e. the manager of Chungar – this is about resolving the Huascarán case] came and expressed the desire to repair things, the general assembly would approve. However, nobody came. They have to admit the error they made. We all make errors but intent is important… It would be good if the mining company in its plans considered workers from the community to generate better relations. Moreover, that they do not commit the same error of deceiving us…Mining generates progress – I defend it, but it has to be responsible. They have to help, not deceive…Relations are bad. They need to be improved – being honest, talking with the assembly. It is their turn to come up with infrastructure plans; they have to leave 10–15% with the community. The *owners* of the company and the community have to negotiate. A representative of the shareholders has to take part in the assembly. It seems that the community relations person is deceiving the company. We need to deal directly with the investors and their representatives who have decision power…They should have a conversation with an extraordinary assembly – that would be beneficial [male community members]…

4.2.7 Status in September 2019

Palma is now an advanced exploration project with polymetallic mineralization of zinc, lead and silver, with some copper. It is a volcanogenic massive sulphide deposit (Volcan Compañía Minera, 2019). It has the potential of a Cerro Lindo or a Coquisiri (well-known metal deposits in Peru) (Fernándes, 2019). The Volcan Compañía Minera web site does not mention conflicts.

4.3 Loma Larga

4.3.1 Field Study Information

I conducted interviews, 14–18 December (inclusive) 2012, in San Gerardo, Chumblín, San Fernando, Girón, Victoria del Portete and Cuenca. The 18 interviewees included presidents and councillors of parish councils, mayors of cantons, officials of drinking water and irrigation systems, farmers, and members of women's organizations, journalists and company personnel. Most of the interviewees were in favour of mining but fortunately, it was possible to meet with some fervent opponents of mining as well. Travel and lodging arrangements were financially supported by the Secretaría Nacional de Educación Superior, Ciencia, Tecnología e Innovación del Ecuador—National Secretariat for Higher Education, Science, Technology and Innovation (SENESCYT) through its programme "Prometeo—Viejos Sabios" (Prometeus—Old Sages).

4.3.2 Project Description

In November 2012, INV Metals bought the Quimsacocha Project from Iamgold, and renamed the project Loma Larga (Long Hill). The project had identified a gold deposit with an indicated mineral resource of 3.3 million ounces of gold. INV Metals is a Canadian company listed on the Toronto Stock Exchange (TSX: INV) with a capitalization of around $30 million. Iamgold suspended operations on the project in 2008, because Ecuador's Constituent Assembly put a temporary stop to mining in the country. The President of the Republic re-established mining under state control and the legislature approved a new Mining Law in January 2009. Iamgold kept its community programme in operation until the sale of the project to INV Metals. The latter resumed exploration activities in March 2013. The Project will minimize environmental impact by constructing an underground mine with underground processing facilities. However, the mine may affect the local water regime (Iamgold Technical Services, 2009).

The project is located in the province of Azuay, Ecuador, at a distance of 30 km. south-east of the city of Cuenca. The definition of the zone of direct influence of the project takes into account environmental criteria (distance from the deposit, hydrographic boundaries and environmental categories) and social criteria (employment and provision of services and infrastructure). Communities in the zone of direct influence are the parish of Chumblín (canton San Fernando, at 8 km from the deposit); the parish of San Gerardo (canton Girón, at 9 km); the communities Gualay and Corralpamba (canton Cuenca, at 12 km), and Portete (at 14 km). Communities in the zone of indirect influence are the cantonal capitals San Fernando and Girón and the parishes of Tarquí and Victoria del Portete (canton Cuenca).

4.3.3 Context

The administrative structure of Ecuador consists of provinces subdivided into cantons, each of which is subdivided into parishes. Each parish contains a number of small communities. The national government is represented in each province by a governor, in each canton by a political chief (jefe político), and in each parish by a political lieutenant (teniente político). The provinces are governed by an elected provincial council headed by a prefect, the cantons by an elected municipal council headed by a mayor and the parishes by an elected parish council (junta parroquial) headed by a president.

The province of Azuay's economic activities included, in order of decreasing importance: manufacturing; transportation, storage and communication; construction; commerce; cattle and related processing. The province's gross total product in 2008 was close to $5.7 billion (Banco Central del Ecuador, 2011). The population of the province of Azuay was 712,127 in 2010, 30% of which were between 0 and 14, 62% between 15 and 65, and 8% over 65 years of age (Instituto Nacional de Estadística y Censos, 2010a). The Loma Larga project interacts with communities in the cantons of Cuenca, Girón and San Fernando. The proportion of the population with unsatisfied basic needs in these cantons was 69%, 75% and 85%, respectively (Méndez Urgiles & Patiño Enríquez, 2013).

In Ecuador, politics strongly influences mining. Some political parties consider mining to be a potential driver of development, whereas others are ferociously opposed to it. There also is considerable popular opposition to mining in certain regions. Some politicians use opposition to mining as a way of attracting votes or power. For example, the prefect of Azuay changed his stance from pro-mining to anti-mining for the Poder Popular movement to put him forward as a precandidate for national President (El Tiempo, 2012). In November 2011, he posited the necessity to "Declare the high plateaus of the country to be free of mining...My spiritual and ideological positions are Quimsacocha free of mining" (Administrador, 2011). In March 2012, he organized a large public event in the city of Cuenca attended by an estimated 30,000 people. On this occasion, he said: "...Here are the men and women who fight for the education of their children, who sacrifice themselves every day to have bread and make a living, here we are to defend the water, here we are the men and women that say to President Correa: NO AGAINST WATER AND AGAINST LIFE" (LibreRed, 2012).

While Alianza País and its leader Rafael Correa won the 2013 presidential elections with 51%, 64% and 51% of the votes, respectively, in the San Fernando, Cuenca and Girón cantons, parties opposed to mining obtained close to 13 and 12% in San Fernando and Girón and only 5% in Cuenca. Therefore, one would anticipate that opposition to mining would be strongest in San Fernando and Girón. Parts of these cantons are indeed opposed to mining. On the other hand, the very strong opposition to mining in the parish of Victoria del Portete in the canton of Cuenca did not translate into a greater share of the vote for the party opposed to mining in this canton, possibly because this parish represents only a small part of the population of the canton.

According to the 2010 census, the total population of the canton of Cuenca was 505,585, while the parishes of Tarqui and Victoria del Portete had 10,490 and 5251 inhabitants, respectively (Instituto Nacional de Estadística y Censos, 2010a). Between the two parishes, they have barely 3% of the total population of the canton.

The Development and Land Use Plan of the canton Cuenca assigned the area of Victoria del Portete and Tarqui to livestock, and some parts as protected areas. The Plan mentioned contamination of the water in the lower parts of the river basins by pesticides, ranching practices, and untreated sewage and other waste. The Plan expected to promote responsible development of mineral resources (Ilustre Municipalidad de Cuenca, 2011: 43).

On 3 October 2011, the irrigation water boards of Tarqui and Victoria del Portete organized a community consultation in which participants answered "yes" or "no" to the question: "do you agree with mining activities in high plateaus and water sources of Quimsacocha?" (Quimsacocha means "three lagoons" in Quechua). Of the 1557 registered voters 1037 participated. 92.8% voted "no" (Defensa de Territorios, 2011). The indigenous organizations Confederación de Naciones Indígenas del Ecuador—Confederation of Indigenous Nations of Ecuador (CONAIE) and Ecuador Runakunapak Rikcharimuy—Confederación de los Pueblos de Nacionalidad Kichua del Ecuador—Confederation of Kichwa Peoples of Ecuador (ECUARUNARI) supported the consultation. The area has seen conflicts related to the Loma Larga project more or less from 2005. Velásquez described and analysed the situation and although her lack of understanding of the natural sciences somewhat affected her analysis, her description illustrated the degree of conflictivity in the area (Velásquez, 2012). The McGill Research Group Investigating Canadian Mining in Latin America also provided a summary of the conflicts surrounding the Loma Larga project (MICLA, 2012). Several interviewees referred to these conflicts.

The cantonal capital Girón and the parish of San Gerardo are located in the indirect and direct area of influence of the Loma Larga project, respectively. San Gerardo is located at only 8 km from the project. According to the 2010 national census, the total population of the canton Girón was 12,607 and that of the capital Girón and of the parish of San Gerardo were 8437 and 1119, respectively (Instituto Nacional de Estadística y Censos, 2010a). Raising livestock was the most important activity in the canton Girón. In July 2010, Lizardo Zhagüi reported that the authorities and inhabitants of the canton Girón declared it free of mining (Zhagüi, 2010). The mayor was opposed to mining because he considered the Loma Larga project to be close to the sources of rivers that supply water to populated zones (No a la Minería, 2011). However, the parish of San Gerardo was strongly in favour of mining. The canton San Fernando consists of the cantonal capital San Fernando and the parish of Chumblín which according to the 2010 national census had 3224 and 749 inhabitants, respectively (Instituto Nacional de Estadística y Censos, 2010a). As in the canton Girón, livestock breeding was the most important economic activity. While the mayor had not publicly pronounced himself either in favour or as opposed to mining, a community interviewee commented that:

> … In general, people of the cantonal capital are opposed, whereas the parish of Chumblín is largely in favour [male community member]…

Another community interviewee commented on the history and functioning of the parishes:

...Parishes were established in 2000 without there being a legal framework or resources, and many have been struggling to advance [male community member]...

An example of the type of challenges parishes face is the potable water system of San Gerardo:

... It was designed 30 years ago for 60 families, while there are 300 families now [male community member, potable water board]...

There also was a comment on the changing context:

...Over the past ten to twelve years there have been many changes. People are more aware of their rights and demand more [male community member]...

Attitudes and leadership were important factors influencing the course of events:

...The two most recent administrations of San Gerardo had very good presidents...For the good of the parish its leaders are preparing for good use of the resources that mining will provide...Well-founded leadership – prepared for what is to come - built capacity for administration [male community member]...

4.3.4   Observed Patterns and Characteristics of Relationships

The government entities that played a role in the relationship patterns in the parishes of San Gerardo, Chumblín and Tarquí include the parishes and their community subunits; their general assemblies, councils and presidents; the parish of Victoria del Portete (because of the fear of attacks it inspired); the cantons of Girón and San Fernando (because of frictions). The provincial prefect and council, the municipality of Cuenca, and the national political lieutenants also influenced the situation. In addition, INV Metals; irrigation and potable water boards; a variety of other organizations, including religious organizations; schools; a commune; Ecuador Estratégico (the national government infrastructure executing agency); and ETAPA, the Public Municipal Telephone, Drinking Water and Sewage Systems Company of the city of Cuenca played a role.

4.3.4.1   San Gerardo

In the parish of San Gerardo, the relationship axes were the parish council, INV Metals and the general assembly. There were many well-linked organizations:

...There is much cross-membership between organizations [male community member, water board]...

Leadership has been important:

>...We have advanced much over the past five years through good management of the president. The two most recent administrations have had very good presidents – the budget was $4.5 million [male community member]...

INV Metals channelled its contributions through the parish council:

>...The parish council negotiates with the company as a representative of the organizations. The latter sometimes interact directly with the company on specific issues...The company has creative mechanisms to finance joint studies for submission to the Housing Ministry - it saved us one year [male community member, water board]...

The relationships landscape was positive:

>...Good and stable relationships both within the parish and with the company – there is respect...The trust in the people that the company shows is a very important factor. Together with transparency, economic support and demonstration of results it has built trust in the community [male community member]...

The relationships involving INV Metals (earlier Iamgold) and its practices developed over a decade:

>...In 2004 the parish council prepared our institutional plan and the general development plan. The company already contracted an excellent technician who acted as the controller [male local authority]...Changes jointly with the company – we signed the first agreement in 2006 that coordinated support and accelerated the process...The continuous presence of the company [even during the interruption of mining activities by the constitutional assembly in 2008] was essential – we did not lose hope [male community members]...

Undoubtedly the panorama will continue to change with circumstances and time, especially when the exploration stage transitions into the mining stage.

4.3.4.2   Chumblín

As in San Gerardo, there were many organizations, and the most important actor groups were the parish council and INV Metals. Since the arrival of the company, several aspects of Chumblín have changed:

>...The company has contributed to driving development...Our personal life has improved...- Women have become more valued and have gained influence, even though God says that men are in command...Machismo has diminished [female community members]...From "victims of mining" we have become "champions of mining"...There has been a strengthening of organizations...In the past things were not good between organizations. Over the past four to five years, there have been discussions within and between organizations through the socialization efforts of the company. It has been a unifying force...Chumblín is much more mining-oriented now [male community members]...

The interviewee comments indicate that the presence of the company and its actions have been a catalyst for internal change in the community, while at the same time leading to a substantial change in attitudes towards mining. The company's community engagement strategy played an important role in this respect: it first established relationships with the commune (that is located very close to the deposit), after that with the women's organizations and from this base with the parish council. The latter now mediates all relationships. The pattern of relationships between the groups of actors appeared to be stable and mostly positive and productive. While the relationship between the company and the community was good, an interviewee clarified that the associations and the community are independent and capable of watching over the company:

…We are not dependent on the company. We know our environment and can monitor the company [male community member]…

One community interviewee mentioned the "minga" as an important mechanism for establishing relationships:

…The minga is an important factor in building relationships [male commune member]…

The minga is the traditional obligation to participate in activities for the common good, such as cleaning the town and organizing the patron saint's festivities. Apparently, the arrival of the company had not affected the minga in Chumblín. Some opponents in Victoria del Portete claimed that mining would cause a loss of local culture, including the minga:

…Even before exploitation they already caused negative impacts during exploration. They destroyed the social fabric and put family members in conflict. By bringing in "Western rationality" they disown our indigenous identity – the minga is disappearing [male members of a group opposed to mining]…

These widely divergent meanings given to the fate of the minga further illustrate the Spanish saying cited earlier: "…nothing in this world is neither true nor false— everything depends on the colour of the glass through which you are looking…". It also reminds us that all respondents were looking through their own coloured glass during the interview. Fortunately, the interactionist framework takes this implicitly into account and postulates credible mechanisms that can change the colour of the glass.

4.3.4.3   Tarqui

In terms of relationship patterns, Tarquí appeared to be similar to San Gerardo and Chumblín, with the parish council and INV Metals being the most important actor groups, with their relationships generally being positive and productive. As there was only one interview in Tarquí, a more detailed discussion is not possible.

4.3.4.4   Other Actors

Additional actors were interviewed in the parishes and cantons opposed to mining (San Fernando, Girón and Victoria del Portete), as well as representatives of organizations opposed to mining. Organizations with which they maintained relations include the Unión de Sistemas Comunitarios de Agua del Azuay (UNAGUA—Union of Community Water Systems of Azuay), the Federación de Organizaciones Indígenas y Campesinas del Azuay (FOA—Federation of Indigenous and Farmer Organizations of Azuay), which forms part of the Confederación de Nacionalidades Indígenas del Ecuador (CONAIE—Confederation of Indigenous Nations of Ecuador). Several interviewees considered the latter very important, as they were of the opinion that the local communities have Cañari nationality. However, according to the national census, the populations of Victoria del Portete and Tarquí identified themselves as mestizos. In a racist environment, statistics based on self-identification of indigenous ethnicity are suspect. In addition, the concept "Cañari nationality" is only applicable to populations that lived in the area before Spanish colonial times. The organizations mentioned above also maintained relations with the parish council of Victoria del Portete. Recent changes led to an increase in the importance of the Confederación de Pueblos de la Nacionalidad Kichwa del Ecuador (ECUARUNARI: Ecuador Runakunapak Rikcharimuy—Confederation of Kichwa Peoples of Ecuador): its new president promised to radicalize resistance (El Universo, 2013).

A community interviewee referred to interactions within and between the cantons and parishes opposed to mining:

> …We have relations with other cantons: Santa Isabel, San Fernando, and Nabón. All are opposed and coordinate their actions. There are other cantons at the national level also [member of cantonal authority]…

The actors in the anti-mining camp kept in close contact and had an interesting take on the meaning they believed INV Metals gave to them:

> …We meet every three or four days for an information exchange and building mutual trust…INV Metals is afraid of us [male community members]…

During one stage of the design of the new constitution of Ecuador, all mining projects were suspended and many indigenous organizations had hoped that this would lead to a prohibition of mining. In the end, the Constituent Assembly decided in favour of mining and promulgated a new Mining Law. This process created a sense of betrayal:

> …There was a national dialogue in 2008 and the constitutional assembly through its mandate #6 (Asamblea Constituyente, República del Ecuador, 2008) suspended most mining activities. Then a new Mining Law was promulgated under the new constitution, without consultation or being informed. We would have liked to see legally binding previous consultations in the law, but that was not approved and now the state can impose mining. This is not an optimal mechanism – it has inequalities [male members of the group opposed to mining]…

Perhaps not surprisingly in view of the divergent meanings of the pro- and anti-mining camps, there had not been much contact between them. However, a careful consideration of reference communities and reading of the material produced by both sides may, with patience, respect and perseverance lead to some bridging of the gaps. The major barrier to overcome is the lack of trust on both sides, and it may be necessary to find and involve intermediaries trusted by both. The existence of political overtones such as the animosity between the anti-mining camp and the national government may make this difficult.

The town of San Fernando (capital of the canton with the same name) has 21 hamlets, with between 50 and 100 inhabitants each. Each hamlet has its president. Budget planning (approximately $6000 per hamlet) is a joint process, per community, carried out in July. Afterwards all meet in an assembly to discuss the overall budget. The capital of the canton appears to be a tightly knit community with converged meanings, and with little contact with the company:

...The canton is small and everyone knows everyone. There is much interaction, which leads to virtue and ethical commitment [convergence of meanings]. I have no contact with the company and do not know much about it...Conflicts are resolved through dialogue [male community members]...

While there was no mineral exploration in the cantonal capital San Fernando or its hamlets, there was some concern about possible repercussions:

...The presence of the company worries me – resistance to mining can create problems. We live in a green zone with water wells all over. If there are excavations, the water will no longer come and the landscape will change. This affects cattle ranching. The future is uncertain. If there is mining there will be fights, and the situation would become uncontrollable. There are much less problems in Chumblín. Everyone chooses his own destiny [male community member]...

Concerns in Girón, the capital of the canton with the same name, focused on water:

...2500 families use the second waterfall for drinking water. Mining will contaminate the River El Chorro. There is no alternative source, and if there were to be a mine there would be a big problem. Where there would not be such a risk of contamination, a mine would be OK. We have to protect the water and not expose the citizens to risk [male community member linked to the local authority]...

According to one interviewee, the company had not done much outreach in Girón:

...Socializing, having meetings, questionnaires, questions – the company has done none of this. They come in without consulting anyone [male community member]...

The canton Girón had anti-mining allies:

...We maintain relations and coordinate with other cantons that are opposed to mining: San Fernando, Santa Isabel, and Nabón. We received support from other cantons at the national level and even from organizations at the international level, with which we have contact a couple of times each year. I do not have any relationship with the company, not because of animosity but because the company has a different purpose [male local authority]...

However, a different interviewee mentioned that the indigenous communities of Nabón were on the verge of supporting mining. In the parish of Victoria del Portete, the potable water and the irrigation organizations that consist of farmers with water use rights and their presidents were considered very important:

…The principal actors are the communities through their potable water and irrigation systems. The water associations are the most important [male community member]…

These water organizations organized a popular consultation process in which a very high percentage of participants voted against mining. Flickr showed pictures of the event (Flickr, 2011). Interviewees did not mention other organizations or associations, although they likely exist.

Communities that were in favour of mining considered Victoria del Portete to be an "anti-mining bastion" that occasionally resorted to violence to make its point:

…When we were sampling water for a baseline study a number of years ago we were ambushed by people from Victoria del Portete who would not let us go and only because police came by were we able to escape. People from San Gerardo on their way to a public mass on top of a hill were ambushed by those from Victoria del Portete who destroyed one of the vehicles of the San Gerardo group because of San Gerardo's support for the Loma Larga project [male community member, San Gerardo] …

Velásquez provided an extensive description of the conflict situation that existed at the time (Velásquez, 2012). However, a local Victoria del Portete interviewee detected the beginning of a change in attitude:

…Attitudes in Victoria del Portete are slowly changing and many are not as opposed as they were before. The risk of me suffering physical harm has decreased [female community member who was in favour of mining]…

In addition, one interviewee commented that the results of a survey supported the contention of the interviewee cited above:

…I have been told that a recent survey by Ecuador Estratégico in the area found significant support for mining [male community member]…

I was not able to obtain a copy of this survey or confirm its existence. It appeared that proponents of mining still feared running into physical risks in Victoria del Portete under certain circumstances and that it was prudent to avoid physical danger. The taxi company that transported me to Victoria del Portete took special precautions to be able to take me out of the building and out of town quickly if need be, by staking out the plaza and having an additional get-away vehicle ready.

Many interviewees commented on the role that relationships played. Examples included the delegation of powers between authorities based on personal relationships:

…He transferred five areas of authority to me [some of which related to spending] [male local authority]…

They also commented on the previously cited improvement of the internal and external functioning of organizations. The latter was due to improvement of the relationships within and between organizations and strong relationships between community members that led to virtue and ethical commitment. In addition, the establishment of trusting relationships with the women's organizations of Chumblín helped strengthen them and led to an increase in productivity for the parish, and to increased respect for women. This also resulted in increased support for mining. According to interviewees not opposed to mining, the community has gained either by improvements in the community or by stability, and the company has gained acceptance. Relationships between proponents and opponents of mining were either absent or adversarial. Protest marches and physical violence were an expression of the latter.

In summary, each parish had extensive networks of cross-member social relationships between organizations and the characteristics of these networks determined if and how they contributed to the functioning of the parish. In each of the parishes of San Gerardo and Chumblín, the arrival of the company had a significant impact on the social networks. In general, the interviewees in San Gerardo, Chumblín and Tarquí considered the resulting changes to have been socially advantageous to their parishes. The processes were dynamic and accompanied by changes in knowledge, perceptions and attitudes, some of them profound. While personal relationships played an important part, there was little dialogue between the pro- and anti-mining camps, although not intentional in all cases. The spectrum of actor groups with which the members of a "camp" maintain relationships (part of their reference communities) was quite different for each camp, and politics was ever present. Chapter 5 provides an extensive discussion of the interactionist processes at work in the various parishes and cantons.

4.3.5 Company Social Responsibility Approach

The company's 15-member social responsibility team developed INV's social responsibility policy and did not base it on a specific existing model. They put emphasis on working *with* the communities. The company maintained offices in each of San Gerardo, Chumblín and Tarqui, as well as the main office in Cuenca. The entities involved in the social responsibility sphere of the company in San Gerardo were two schools; the communities of Cauquil, Santa Ana, Crital Aguarongos, San Gerardo Centro, San Martín Grande, San Martín Chico, and Bestión; INV Metals; Seguro Social Campesino (Rural Social Security); the police of San Gerardo; the fire brigade of San Gerardo; and two associations.

INV Metals designed and implemented all its CSR initiatives together with the parish councils. Both the company and the parish councils contributed financing and, depending on the case, the company provided or contracted technicians. The initiatives were part of the parish development plan. Examples include support for the establishment of women's organizations, construction of irrigation canals, planting trees and looking after the environment. In addition, the company contributed in health, education, roadwork, culture, social matters, production and capacity building (often the company provided experts). It also supported "Buen Vivir" ["Buen Vivir" is a concept in Article 275 of the Constitution of Ecuador that links to rights, and to exercising responsibilities linked to living together in an intercultural context and in harmony with nature (Secretaría Nacional de Planificación y Desarrollo Ecuador,

2009)]. The following interviewee comments indicate that the interactions between the company and the communities that are in favour of mining appeared to be well structured and understood, and productive:

> …The general assembly puts forward development proposals that are discussed with the company…The national government contributed $80,000 and the company $97,000 for the projects in this year's budget…The company is ready to cooperate in development…there is accountability and transparency involving the population, local government and the company…We do everything with them…The agreement between the company and the community contains an execution plan. …The plan is managed by the company, as this avoids bureaucracy and is a fast mechanism…The trust of the company in the people is a very important factor [as mentioned earlier]…The objectives of the company are clear: support and promote productive activities… The company's manager designs the social responsibility policy. The program is developed through the parish council and the president is a key person…There has been capacity building with the company in several areas – they have money…I have read the company reports on the activities and I agree with what they say [multiple comments]…Yes, the company is concerned about the people in the region [multiple replies]…They pay attention to the elderly and the sick…The company offers honesty and attention to all…We are building capacity in communications and citizen oversight…The approach has been designed by the Ecuadorian team – the company contributed much. …It is important to ensure that the employees do not become disconnected from their community – we are looking for programs that compensate for the difference in income such as productive activities and the local provision of services and supplies to the company …The social responsibility manager looks after the policy, the team executes the programs all of which are carried out jointly with the communities… [male company employee].

The interviewee comments paint a picture of a well-managed CSR approach by both the company and the communities in favour of mining. While the philosophical and moral dimensions of social responsibility should be a primary driver, from a pragmatic perspective considerations of return on investment also play a role. The investment in a team of 15 experts and in contributing to a number of communities may be of the order of $1 million annually. It is difficult if not impossible to calculate the return on such investment as it is not easy to measure social matters in dollars. However, a paper by Davis and Franks suggests that the cost of not investing in social responsibility can be very high indeed (Davis & Franks, 2011). The case of Meridian Resources, which lost its Esquel projects through a series of avoidable social responsibility mishaps and had to write down more than $600 million, illustrates this well (WRI: Development Without Conflict, 2007).

Interviewees opposed to mining belittled the company's social responsibility approach, comparing it to "beads and mirrors", and were suspicious of the company's intentions

> …They are lying. They give little gifts to the community leaders, and small public works. They should go away…The communities are building capacity themselves through events, exchanges and travel…One of us traveled to Switzerland (for the UN), Colombia and Peru to see mining in action…I read the company reports carefully and found them to be almost imaginary. They have nothing to do with reality…I do not know much about the company…We know – donations of chickens, guinea pigs, communal houses, etcetera. Slaps in the face to the people….They are not at all concerned about the people – no international company is. They co-opt the people by sending them to Canada or by signing agreements with

the universities [Loma Larga/IAMGOLD and the University of Cuenca signed an agreement for the study of the hydrology of the zone in which the project is located] [male members of the group opposed to mining]

The meanings given to the company's social responsibility approach by the people in favour of mining were quite different from those given by the opponents of mining. It is interesting to note that some of those who are now pro-mining used to be anti-mining. They changed the meanings they had through interactions with their fellow citizen bridge builders. In Chumblín, the bridge builders were the commune and the women's organizations, and the change spread out from there. They illustrated the (literally translated) Dutch saying, "when the first sheep has crossed the dam, more will follow".

4.3.6 Perceived Present and Future Benefits and Harms

4.3.6.1 Benefits

Benefits mentioned by interviewees included relationships; employment; education; capacity building; knowledge; a stronger community; and, very importantly, hope for the future:

> …Personally I have gained many good relationships and in my area of responsibility I have benefited much. However, financially I lost [by investing so much time and effort in the parish]…The entire population has benefited: investment in schools; employment of 30 people in the company and in public works; agriculture and livestock capacity building; a bakery; crafts; talent development; a stronger community that organizes itself; and knowledge – it is important to maintain the organizations[male community members]…The increased employment has reduced emigrations rates which has been good for families… The company has won over the people…The most important aspects are economics (hope) and well-founded leadership that is prepared for what is coming…We have all gained… I would like to see the company start production so that there will be employment, a better increase in infrastructure and a change of life: less anguish about food and other things. It would be a disaster if this did not happen – may it accelerate [female community members]… Yes the company has the support of the community…Our personal life has improved The company has not yet benefited as they are not yet into exploitation…When the mine comes there will be more employment – there are some concerns about contamination…There has been organizational strengthening…The company will benefit in the future [male community members]…Women have gained self-respect and influence… [female community members]…

These comments show perceived present and future benefits in the social; the economy (employment, income); infrastructure; and human capital formation. "The social" (built on relationships and interactionist processes) was the engine that drove everything else: organizations strengthened and as a result could take on development projects with greater self-confidence and better governance—this included involvement in planning and execution of the infrastructure projects. Social and human capital building go hand in hand: with greater skills come more productive and enjoyable interactions.

4.3.6.2   Harms

Opponents of mining mentioned a number of downsides (some of which were mentioned earlier) that included potential conflicts and divisions; corruption; contamination; loss of culture; land grab; and criminalization of protest. An advantage they saw was that those opposed to mining united in resistance:

> …There hasn't been any benefit and exploitation will lead to conflicts…The future is uncertain and worrisome…They have not socialized. There have been many protests and marches… I am worried about the future for the water, for contamination. We will lose our identity and culture, and it would be catastrophic. We put up resistance to the government. We claim that it [the situation] is unconstitutional. According to the constitution the National Assembly could not do what it did. We had a first hearing and the judges will pay a visit to the project…Pillage by Canadian companies… They destroyed the social fabric and caused conflicts within families. In terms of culture, the "progress" of modernity they bring repudiates the people's indigenous identities: the minga is disappearing. Western rationality affects the spiritual. Economically, they want to snatch our communal lands in the páramos [moors] by fraud, declaring them barren lands. They have affected institutions (co-optation): they paid the University of Cuenca to carry out studies in the páramos, which came up with information in favour of the project through the Programa de Agua y Suelos (Water and Soils Program).…We are afraid that the government is criminalizing protest – they are already doing it using the army as an instrument.…The city of Cuenca is the third largest city in Ecuador and a national heritage site: if they begin exploitation Cuenca will be finished. Water is life: "no se vende, se defiende" (it is not sold, it is defended)…There have been no benefits, but there has been one advantage: they have lifted our spirits. Over two years we have had mobilizations, resistance and detentions (more than 150 people in one night) and three people have been sentenced [male members of the group opposed to mining]…

The opponents of mining tended to be wordier in their replies and refer to the bigger picture that included the national government, the constitution, culture and water as a sacred symbol. In interactionist terms, this relates to the types of interactions between peers in this group of actors. As mentioned by one of the interviewees, they meet every three to four days and have produced pamphlets on the constitution and on water (Guartambel, 2012). These actions made meanings converge through the teacher–student, leader–follower, organizer-organized role and counter-role identities and thereby made group action against the project possible (Deaux & Martin, 2003). In any society, different opinions exist on how to manage natural resources. Ideally, transparent dialogue and negotiations lead to an agreement. The availability and sharing of pertinent, high-quality information play an important role in this respect. Unfortunately, there is a great risk of the dissemination of incorrect information, or information that is misunderstood or misinterpreted. An additional challenge is the low level of literacy that may strongly influence communication.

4.3.7   Status in September 2019

A plebiscite in the canton of Girón declared it closed to mining and an application for a proposed referendum on mining activities in the province of Azuay is under way.

The outcome is uncertain (Junior Mining Network, 2019; Mining Journal, 2019; Mining Watch, 2019). While INV Metals established excellent relationships with some communities, other communities never accepted mining and dialogue appeared impossible.

4.4 Fruta Del Norte

4.4.1 Field Study Information

The interviews took place between 19 and 22 February 2013. We interviewed 20 people, including presidents and councillors of the parish council and of communities, farmers, leaders and members of associations, educators, entrepreneurs, journalists and company personnel. Unfortunately, it was not possible to talk with strong opponents of mining, as they were not available during this period. The study team consisted of Jésssica Marcayata and Dayana Velasco of the Instituto Nacional de Investigación Geológico, Minero y Metalúrgico del Ecuador (INIGEMM) and me, and Camilo Agua was the driver of the INIGEMM vehicle. INIGEMM also provided logistics support. I received financial support from the Secretaría Nacional de Educación Superior, Ciencia, Tecnología e Innovación del Ecuador (SENESCYT) through its programme "Prometeo—Viejos Sabios (Prometheus—Old Sages).

4.4.2 Project Description

The project is located in the Cordillera del Condor in the province of Zamora Chinchipe, about 195 km east-northeast of the city of Loja, close to the parish of Los Encuentros that is part of the canton Yantzaza. There has been artisanal and small-scale mining in the area for a long time. Various companies carried out mineral exploration activities in the area between 1985 and 2005, Aurelian Ecuador being the latest. A geological reinterpretation in 2006 led to the discovery of an important ore body, and in 2008, Kinross Gold Corp acquired Aurelian Ecuador. The ore body has indicated resources of almost 16 million tons of ore with a grade of 11.2 g of gold and 14.3 g of silver per ton, which translates into close to six million ounces of gold and seven million ounces of silver (Henderson, 2009). Kinross extended and adjusted the corporate social responsibility programme that Aurelian Ecuador had started.

4.4.3   Context

4.4.3.1   Province of Zamora Chinchipe

The gross value added of the economy of Zamora Chinchipe in 2007 was US$118 million in 2007 (Banco Central del Ecuador, 2011). Major sectors of the provincial economy were, in decreasing order of importance: public administration; public administration, defence, obligatory social security plans; construction; education; agriculture and related activities; transportation, storage and communications; wholesale and retail; mines and quarries. The total population of Zamora Chinchipe was 91,376 in 2010, 38% of which was between zero and 14 years of age, 57% between 15 and 65 years of age, and 5% was older than 65 years (Instituto Nacional de Estadística y Censos, 2010a). The population of canton Yantzaza was 18,675, distributed between the parishes of Yantzaza (12,356), Los Encuentros (3658) and Chicaña (2661). Its distribution between age groups was 0–14 years: 38.9%; 15–64 years: 56.4%; and over 65 years: 4.7%. 73% of both the total population of the province of Zamora Chinchipe and of the canton Yantzaza had Unsatisfied Basic Needs in 2009 (Instituto Nacional de Estadística y Censos, 2010b).

In the 2013 presidential elections, more than 60% of the votes went to parties that were in favour of mining. This could mean that more than half the population was also in favour of mining, an interesting context for the relationships between actors and groups of actors touched by the Fruta del Norte project. However, mining is only one of many reasons that would lead a person to cast a vote for or against one particular person or party or another Also, Salvador Quishpe, the prefect of Zamora Chinchipe, was a strong opponent of mining, as was Dr. Angel Erreyes, the mayor of Yantzaza canton. Clearly, mining affected some of the political differences in the province, and national government authorities in Zamora Chinchipe interpreted the election results entirely through a mining lens. The canton Yantzaza like many others in Ecuador is young: it was established in 1981. Only the Shuar people have lived in the area for a long time. Apart from them, there are no other ethnic groups with ancestral history in the Ecuadorian Amazon region. The region was colonized by settlers in the 1960s, and its economy has largely depended on artisanal gold mining, although there is also agriculture and ranching (Agreda Orellana, 2011). The parish councils became consolidated beginning in 2000, and the 2008 constitution converted them into Gobiernos Autónomos Descentralizados— GAD (Autonomous Decentralized Governments). The parish council elected for the 2009–2014 period consisted of one member and one substitute from each of the political parties Alianza País, Socialistas, Movimiento Acción y Servicio, Sociedad Patriótico and Pachakutik (indigenous). The President of the republic belongs to Alianza País. In view of the considerable political differences between the parties, it is not surprising that the Plan de Desarrollo Y Ordenamiento Territorial de Los Encuentros—PDYOTLE (Development and Land Use Plan of Los Encuentros) identified challenges in political-administrative organization (Gobierno Parroquial de Los Encuentros, 2011). According to the PDYOTLE, agro-productive activity is at a subsistence level, forcing people to seek other sources of income such as artisanal mining. Of the local population, 11% is Shuar, 5.3% is Saraguro and the remainder is mestizo.

4.4.4 Observed Patterns and Characteristics of Relationships

The groups of government actors that played a role in the relationship patterns in the project area include the provincial prefecture; the mayor's office (of the canton Yantzaza); the parish council, the communities of Los Encuentros (subunits of the parish), the national government and its political lieutenant and the governor. Additional actors included the "community in a general sense", political parties, the water council, the Shuar Federation, the Catholic Church, Kinross, Kinross' employee association, the artisanal miners association, the pick-up truck cooperative, the women's organization, APEOSAE, the cattle association, the volunteer group, the sports clubs, the cultural commission, the TV station; and educational establishments. The relationships matrix showed that most relationships were positive and that the parish council and Kinross were the axes around which most relationships converged. While some would say that the perceived importance of Kinross is a natural consequence of the focus of the study, I designed the first two questions of the interview guide to elicit information on all actor groups and the patterns of responses consistently pointed to Kinross and the parish council. The relationship indicators and Chap. 5 will discuss the interactionist processes for this case in detail.

4.4.5 Company Social Responsibility Approach

Kinross played a key role in the development of the Código de Comportamiento Ético (Code of Ethical Behaviour) of the Consejo de Minería Responsable del Ecuador (Responsible Mining Council of Ecuador) (Ekos, 2012). It based its CSR approach on its values and developed it against well-defined criteria:

> …The company bases its work on four principal values: Putting People First; Outstanding Corporate Citizenship; High Performance Culture; and Rigorous Financial Discipline…We also respect communal living …Criteria against which we developed the local social responsibility program included sustainability; alignment with the Plan de Desarrollo y Ordenamiento Territorial de Los Encuentros—PDYOTLE (Development and Land Use Plan of Los Encuentros); added value; local development; and support and strengthening of social organizations (one of the reasons for which we coordinate all actions through the parish council)…Our planning is based on the company's values; applicable legislation; environmental management plan (a sub-plan of which is the community development support Plan); and international standards. The budget is approved at Kinross' head office in Canada, but designed and decided upon locally. The details are developed jointly by local personnel and the community – they share responsibilities and budgets [male and female company personnel during a group presentation]….

Each Gobierno Autónomo Descentralizado—GAD (Autonomous Decentralized Government) is required to develop its development and land use plan. Kinross supported the parish council so they could hire a consultant to help Los Encuentros with the preparation of its Plan de Desarrollo y Ordenamiento Territorial de Los Encuentros (PDYOTLE—Gobierno Parroquial de Los Encuentros, 2011). It covers a period of 20 years and is reviewed annually. Local plans are coordinated with regional plans, which in turn are coordinated with the national plan. The Secretaría Nacional de Planificación y Desarrollo—SENPLADES (National Secretariat for Planning and Development) oversees the process. The interviewee comments reproduced below show that community members perceive Kinross' CSR approach as being a joint effort with the community that follows well-developed protocols for interaction:

> …We do it [design of the social responsibility strategy] together, Kinross shares technicians. The zones [in which the work will be done] are selected by mutual agreement. Kinross has a well-organized social responsibility system. Each community [sub-unit of the parish] has its leader who interacts with the company…There are meetings in which the entire community dialogues and afterwards a formal request is made through the executive. We start a dialogue when we need to. Kinross is nimble, not much bureaucracy is generated…Community needs are explained to the company every month. The community has many needs, reason for which petitions are presented often. It is important to mention that there still is some lack of company attention to the community…

According to several interviewees, Kinross and the community were involved in extensive building capacity projects:

> …Yes, the community is building capacity in community organization, leadership, business projects in the community, productive topics, cattle ranching, food preparation, defensive driving, welding, electricity carpentry; and industrial arts … The company contracts technicians for this… The community is building capacity with the help of the company, For example the program of condensed basic education in cooperation with the Ministry of Education. Kinross needs Ecuadorians as mining technicians, and this takes ten years of education. We need a system to prepare ourselves. 170 people participated in the condensed basic education program, 50 of whom were Kinross employees. Many continued to study afterwards on their own hook…I know about the training provided to APEOSAE (Association of Small Organic Farming Exporters of Southern Ecuadorian Amazonia) [male community members]…

However, several community members were not aware of these initiatives:

> …I do not know if members of the community are receiving training [female community member]…Right now the community is not building capacity…The community is not receiving any training [male community members]…

The discrepancy between observations made by different groups of interviewees illustrates that, even though Los Encuentros is not a large parish, it consists of networks not well connected to each other. The fact that the population of Los Encuentros is scattered over a number of subunits may contribute to this. From an interactionist perspective, this reduces the number of interactions between groups of actors and slows down the convergence of meanings. Interviewee comments cited in the interactionist analysis of this case study in Chap. 5 show similar discrepancies with respect to perceptions about Kinross' versus Aurelian's relationship with the communities [subunits]. Kinross contracted experts to deliver these courses and it provided didactic materials, paint, classrooms and scholarships. The company planned its programmes based on needs and dialogue with the community, and paid much attention to internal capacity building:

> …Everything depends on the need. We need to prioritize what we teach. The community is beginning the development of a participative budget. They understand it – it is a process of sharing and to reach a balanced development [male employee]…What we [company staff] teach depends on the objectives of our job, for example, someone has given a course on CSR to journalists, which also enhanced the reputation of the company and led to links with the press. We receive annual training for example on teamwork that led to results: we now have self-managed teams. An external firm [female employee] provided the training… Our internal program addresses three areas: technical training, condensed basic education (as part of the program mentioned below), and values, environment and health and safety. Employees obtain expertise certificates that help increase their opportunities for future employment [male employee]…

4.4.6 Perceived Present and Future Benefits and Harms

In addition to the contributions mentioned in the previous section, interviewees considered the provision of employment as a very important contribution:

> …With respect to the percentage participation of the company in areas important to the community the outcome is: employment 30% and education 70%… The arrival of Kinross was very beneficial to the community through the opportunity to create a source of employment…It generated work [male community members]…The company supports the community through employment and also public works (for example the construction of bathroom facilities). But there has to be much pressure from community representatives to achieve the objectives [female community member]…

Other contributions to the economy that interviewees mentioned include support to an organic grower's organization, Shuar culture, parish council projects and the local economy:

> …APEOSAE has seen major development over the past number of years and is even exporting its products, which is a great achievement for the organization, especially in terms of the quality standards it reached… The company certainly makes an economic contribution through work, and also has provided Shuar traditional costumes…The company contributes $75,000 to the parish's annual budget [for joint projects]…Many individuals and groups have been able to do things they could not have done before (for example the hotel, the half-ton truck cooperative, businesses [male community members]…

The local press also mentioned some of these contributions (Arias, 2013).

Kinross contributed specific support to the Shuar Federation in infrastructure, training, services and culture. The Shuar Federation found accompaniment to be much more important than financial support. Non-Shuar community members commented that they are now prepared for a mine and pointed out that the Fruta del Norte project led the national government to invest heavily in infrastructure in the area through its company Ecuador Estratégico. It was interesting to note that various interviewees mentioned social responsibility and its importance. The following comments illustrate the above

> …The agreement between Kinross and the Shuar Federation led to improvements in infrastructure, capacity building and the establishment of services. In this sense they are following the Canadian model…Their money is not where the value is – accompaniment is what we need. The plans of the Shuar federation were developed with Kinross' accompaniment [male Shuar community member]…Yes, the company is concerned about the people in the area [multiple comments by male and female community members]…The ore body exists [and has been defined] and even if there is no contract [between Kinross and the government] the community is prepared for a mine [male community members]… So far, the town has benefited. Thanks to the presence of the company, the government undertook many public works for millions of dollars. It has made an enormous difference…In addition to the royalties that Kinross pays, it supports the community in what it needs, added value to the work they are doing [male community members]…There is a good future for the community as regards social responsibility…With respect to social responsibility it is important to try and maintain it even if the company leaves it would be much needed because all of us in the community need a push to carry on [female community members]…People have changed because they saw the change – with the support to employment [male community members]…

Reported disadvantages included the effect of the Kinross' wages on purchases by its employees and the risk of abandoning farming to take on insecure jobs:

> …In the market they sell products at a higher price to Kinross employees because the vendors assume that they have much money because they work for Kinross. The government should put in price controls…Personally I have not received any benefit [male company employee]…The arrival of the company was damaging to the community…In the past we lived well on only agriculture, after mining people started going to that work and left agriculture and now they are in a situation of uncertain employment…I have worked for the company and I became aware of the injustices of mining [male community members]…

Kinross regularly contributed to social events such as the annual parish celebrations, summer camps for the children of the parish, motivational presentations, family activities and social development programmes. Kinross employed various forms of communicating inputs and outputs such as a community newspaper (that was being read); monitoring of perceptions; radio; television and posted notices. The comments cited above indicated that respondents believed that the community had benefited from the presence of the company in a variety of ways that include employment, training, knowledge, organization, infrastructure (thanks also to related investments by the national government), economy, retention of Shuar culture. As a result, the community was in a better position to face the future. Some interviewees expressed minor concerns. Many interviewees were of the opinion that the company had also benefited because it had gained the acceptance of the people, established friendship bonds and was able to carry out its work, and because it had access to productive, relatively cheap labour. It has access to a valuable gold resource and will benefit when the mine opens:

> …Kinross obtained its social licence and it has the trust of the people. It has productive workers from the community…Yes the company benefited because it has cheap labour, which is in its interest… The community now knows the Canadian company and therefore supports

it…I do not know if the company has benefited but I believe that one of the benefits would be striking up friendship bonds and be united with the community [female community members…Yes the company benefits because it will extract the minerals…The entire community changed and accepted the company because of the source of employment it provided… I believe that yes, the company has benefited because it will extract gold from the area…So far the company has not gained but in the future it will gain economically. In other terms, yes, it has gained [trust, respect]…The company gained because of the appreciation and the gratitude of the people [male community members]…

The lack of progress in the negotiation of a contract between Kinross and the national government frustrated several interviewees:

…I believe that the company has not yet had any benefit because the contract [with the government] has not yet been signed…The company has lost so far: the contract has not yet been signed…Everyone is losing by the delay in the signing of the contract between Kinross and the government (the community, the government and the company)…While there is no contract we are all losing: the company, the community and the government [male community members]…

There was fear that, if Kinross and the government could not come to an agreement, a Chinese company might buy the project, and Chinese companies have a bad reputation in Ecuador:

…I am not opposed to Kinross because they take advantage of natural resources. We know that the Chinese have neither environmental nor social responsibility…The Chinese companies only take out the minerals, they bring their own technicians and do not employ local people…We will not permit the entry of Chinese because we know that they do not treat their workers well, there are comments from friends and everyone that the Chinese are no good…In Panguí there was strong opposition to the exploration project and the company sold it to the Chinese [EcuaCorriente S.A.]. Now there is no popular support left. The company that bought the project has a different philosophy – they consider the public works carried out [by Ecuador Estratégico] using advanced royalties to be sufficient and they do not have a special fund for community development such as Kinross has. If the government does not sign with Kinross, people are very afraid of the Chinese. This would cause grave dangers to the soil, the environment and culture…The community will not allow the entry of a company other than Kinross [male community members]…

The company had gained the trust and the support of the community, access to productive workers at reasonable salaries in Los Encuentros and access to a valuable gold deposit. It also made good progress towards its goal of community development.

The observed course of social events and the perceived many benefits and few harms resulted from the interactionist processes that Chap. 5 will discuss. The Fruta del Norte case well demonstrates the processes represented in Fig. 3.2, and the productive outcomes suggest that the company and the community based their meanings, interpretations, decisions and changes on valuable underlying principles that include dialogue, meeting transactional needs and tactical dimensions of building trust. Chapter 6 will present a generalized model that builds on the case studies. Kinross and the Ecuadorian government could not reach an agreement and Kinross ceased work on the project. The government of Ecuador commissioned a study on its competitiveness in attracting foreign direct mining investment that resulted in some changes to its policies. Because of these changes, there was renewed interest in investing in Ecuador and in October 2014, Lundin, a Canadian company with a good business and social reputation, bought the project. The parish of Los Encuentros presented Kinross with a plaque expressing its appreciation, and symbolically handed the keys of Los Encuentros over to the CEO of Lundin (personal communication, Dominic M. DeR. Channer).

4.4.7   Status in September 2019

Fruta del Norte has begun mining of its first production stope. In 2017, it received an award in the UN Sustainable Development Goal 8 (Decent Work and Economic Growth) category (Lundin Gold, 2019a, 2019b).

4.5   Cauchari-Olaroz

4.5.1   Field Study Information

The study team consisted of Bernarda Elizalde (then CSR Manager of the PDAC) and me. The interviews took place between 24 and 28 October 2011. We interviewed 16 company employees, six community members, one provincial authority, one member of the provincial legislature, one business organization, three academics, one trade union representative and one member of a religious organization. Ten of the interviewees were female and 21 were male.

Many of the interviews were set up in advance with the help of Graciela Medardi, Director of the School of Mines, National University of Jujuy.

4.5.2   Project Description

Minera Exar was a fully owned subsidiary of Lithium Americas (TSX: LAC, OTCQX: LHMAF). At the time of the case study, major shareholders were Mitsubishi Corporation and Magna International. Its market capitalization was of the order of US$115 million. The company focused on the exploration for and the production of lithium and potassium from brines found underneath extensive salt flats in the Argentine puna (high plains). The Cauchari-Olaroz project will recover Li from the two salt flats with the same names that run north–south and cover an area of roughly 100 km by 10 km. The company's main concessions are Cauchari-Olaroz, Inca Huasi, Pocitos and Arizaro. Its Cauchari-Olaroz property was at the prefeasibility stage and the company was identifying production parameters using a small pilot plant that was located at the edge of the salt flat, at 70 km from the company office in the town of Susques, Department of Susques, province of Jujuy, Argentina. The project is located at an altitude of about 3950 m above sea level. Minera Exar had the rights to subsurface minerals (Li in this case), whereas other types of companies have the rights to surface minerals such as borates. There are about 20–23 "borateras" (borate producers) in the area, many of them quite small, which have been producing borate for ten to 20 years.

4.5.3 Context

The province of Jujuy covers an area of some 53,000 km^2. The puna, a high plain at an altitude more than 3500 m above sea level, takes up the major part of the province and extends across its north-west part. The leaching and transport of elements from volcanic strata formed the salt flats and brines characteristic of the desert landscape. Evaporation of the water contained in the leachates led to the precipitation of the salts. They contain economically valuable elements such a boron, potassium and lithium. The high altitude and associated scarcity of oxygen, lack of precipitation and near absence of hydrography, thin plant cover, and hard climate with extreme variations in temperature explain the low population density. There has been a reactivation of mineral exploration in the province since 1932. Jujuy has been a mining province since pre-Inca times (Alonso, 2010).

According to the 2010 census, Jujuy had a population of 672,260, of which 3757 lived in the Department of Susques. Its population was relatively young: 27% was younger than 15, and 64% had ages between 15 and 65 years (Instituto Nacional de Estadística y Censos (INDEC), 2010a). For the principal occupations in urban zones, the average monthly remuneration in Jujuy was about 75% of that for Argentina as a whole. The monthly remuneration of 10% of the working population exceeded US$3000, and 50% of the population had a monthly remuneration of less than US$1500 (Instituto Nacional de Estadística y Censos (INDEC), 2010b).

Over the period 1980–2002, the average proportion of the population of Jujuy that was economically active was only 34%, as compared to the national average of 42.3% (Martínez, Golovanevsky, & Medina, 2010). 26.1% of households in Jujuy had unmet basic needs. For Susques, the town in which Minera Exar had its office, the corresponding figure was 42.0% (Instituto Nacional de Estadística y Censos (INDEC), 2001a). 73.1% of the population of Jujuy was living below the poverty line, and 36.4% was living below the extreme poverty line. The corresponding numbers for Argentina as a whole were 57.5 and 27.5, respectively (Instituto Nacional de Estadística y Censos (INDEC), 2001b). In the puna region of Jujuy, there is a significant concentration of indigenous communities and the total number of indigenous people is around 41,000 (Bidaseca, 2011).

Jujuy was the province most advanced in transferring land titles to indigenous peoples, while the indigenous communities at the same time played a strong role in the debate together with officials and legislators (Borghini, 2010). As a result, most of the land in the department of Susques has been "measured and transferred" to indigenous communities. The claims of the Cauchari-Olaroz project are located on aboriginal lands that belong to the communities of Puesto Sey—Termas del Tuzgle; Pastos Chicos—Manantiales; Huáncar; Olaroz Chico and to a lesser degree Susques—Pórtico de los Andes and Catua. The most important settlements in the department are Susques, Puesto Sey, Pastos Chicos, Huáncar and Olaroz Chico (Sarudiansky, 2008). The communities in the area attach an important meaning to the land:

> …The communities in this region are very much attached to the land [female company employee] …

4.5.4 Observed Patterns and Characteristics of Relationships

Several interviewees indicated that face-to-face interactions are important, whether they are positive or negative:

> …The people here [in this area] are very quiet and in general look after me very well. It is easy to work with the people from here. They hard-working and do not have problems with anyone…There is very little conflict: issues are always clarified through conversations … Living together [in the camp] creates a special climate – it is not perfect … We did not work well together, we had to ask for permission too often, there was much affection but also much tension. In the end, the person with whom it was difficult to relate was let go …A third party could extract bad opinions and good opinions [from employees when there is an issue] to be able to resolve the issue. Maybe the new person in Human Resources could help change the situation. Every now and then a person from outside could analyze the situation [male employees]…

The community relations team handled most of the direct interactions between the company and the communities. They put great effort into building personal relationships with the communities. Sometimes this created ethical dilemmas for them:

> …There is a continuous go-around between community relations, the communities and the CEO, with much social dynamics. We take part in the general assemblies of all the communities and get to know the people very well…For issues related to the communities, the other parts of the company always contact us…Personal friendship with community members can cause ethical dilemmas. For example, sometimes we are entrusted with personal information. How to maintain work and personal matters separate? Keep within the company what needs to stay within, and make sure to keep outside of certain comments by community members – sometimes we cannot react [female employees]…

The management of the company considered CSR to be very important as indicated by the comments below, from the perspectives of the communities' well-being and of operational and relationship considerations

> …The company has the potential to change the local communities' life and make sure that those changes are positive, and the negative risks mitigated… CSR has to do with the human side of things, and how are you going to manage it. Take care of the changes that are to come due to the mining activities. Some decisions in this regard cost money, and we allocate it

through our community relations department… Is it spending money or investment? CSR is an investment because if you do not allocate resources to these matters, the communities will not allow you to operate…It is important to be present. Small things are important, a consistent relationship and contact with locals is essential…The community relations staff reports directly to the CEO because there are decisions that need immediate and direct attention. The budget managed by the community relations team is independent from the rest of the company's budget [male company CEO]…

The network of relationships was still fragile and dependent on a few critical nodes: the community relations team, the company CEO and specific community members that had spearheaded the establishment of relationships with the company. As was mentioned in the discussion of the theoretical framework, the processes were dynamic, and this study provided only a snapshot. Changes in key personnel such as the CEO and the community relations team could have a major impact because of a change in philosophy or in the continually changing dynamics of interpersonal interactions. The communities were well aware of the risks of possible changes in company personnel or approach:

…For now they respect us, but afterwards we do not know…The respect for the local people has to do with the personnel of the company. Let's hope that if there is a personnel change the same level of respect is maintained…We are concerned and we want to know what will happen in the future when they grow, when they begin to produce lithium [male community members]…They are good in the social, but with time we have expectations, especially when the company begins to produce lithium…The ideas have to come from us, because they have to be for the long term. Because one day the company will leave. We know that the state has to participate too for things that we cannot do [female community members]…

From engineering or classical organizational behaviour perspectives, such situations present a challenge, but they are par for the course in symbolic interactionism. Its flexible framework takes into account continuously changing meanings, interpretations, reference communities and decisions and facilitates the identification of opportunities and threats for all the actors involved.

As it was possible to interview a fair number of both community members and company employees, it was possible to compare "within company" and "company-community" relationships in terms of the relationship indicators. I demonstrated the process I used to arrive at the summary descriptions for the relationship indicators in this chapter for Fruta del Norte case. In the following section, I discuss the "within company" and "company-community" relationship indicators in groups of two.

4.5.4.1 Trust and Respect

Within the company

The comments below suggest that there was trust and respect within the company

…The relations between managers are good and there is much dialogue…Management is flexible and supports staff…There is much freedom of decision and of organization of work across the company… Limits are established appropriately and the company is well organized – it delegates to the various areas of work and has good practices… I am proud to work for a company that wants to do things well, gives people the opportunity to do things well, and has confidence in us… Discrimination is within a person and there are no problems – there is respect within the company…. Also higher level employees trust lower level employees and let them work freely [male employees]…

Community-company

While the community member comments do not specifically mention the words "trust" and "respect", those reproduced below implicitly recognize the presence of trust and respect:

> …Exar is the only company that has come to talk with us directly from the beginning [female community member]…They always participate with all the people in the town. People like it that they come because it helps us…In the community everything is all right with the company because they have consulted us from the beginning and they always communicate changes and through agreements…I have seen that the company does not impose its ideas…The company is doing things right, helping the community including doing what they are not required to by law, this makes us feel that they are good neighbours [male community members]…

4.5.4.2 Communication and Mutual Understanding

Within the company

While communication within the company was not bad, improvements would be possible in cross-area linkages such as between community relations and other departments to avoid "crisis management". Employees also would like to receive more information about plant operations and mitigation approaches in case of a problem:

> …Electronic media are used extensively… Agreements are confirmed in writing and improvements are made as necessary… Cross-area linkages could be improved and some technical problems avoided….The community relations department could work in a more coordinated and continuous manner with operations and other departments to avoid that its work turns into "extinguishing fires" instead of creating synergy…In the company they always talk about our responsibilities. We hope that it disseminates this more with examples and workshops on human resources and rights and obligations…We need them to give us more explanations about mitigation measures [for when something goes wrong in the plant]. It would be ideal if they could explain what the plant will be like [male employees]…

Community-company

There was good mutual understanding between the company and the communities:

> …They inform us of the plans and we want them to proceed…They respond to our concerns, and well…The company always consults us through the assemblies, and because of this we have good relationships…The important thing is to speak truth – they even informed us about moving a bit of soil…We would like them to come and talk to us about risks and mitigation every three months [male community members]…

4.5.4.3 Conflict Resolution

Within company

While there was a low level of conflict within the company, the decision of a local employee to seek redress through his community's leader rather than go through the company structures suggests that the establishment of a more formal conflict resolution mechanism would merit consideration. At the same time, the fact that the community leader immediately approached management and that the problem got resolved speaks well for communication between the company and the community. However, the comments cited below suggest that there have not been serious conflicts within the company, or between the company and the communities:

> …There is a low level of conflict and most problems get sorted out through conversations… Neither Minera Exar nor Sales de Jujuy has an explicit grievance mechanism and both handle complaints through their Community Relations departments [external interviewee in the provincial capital]…Complaints are channelled through Management and Human Resources. Community Relations deals with external complaints and the reports go to the manager of the Susques office who makes the decisions. The CEO meets with community leaders to respond to the community…When formalized, the grievance system could be improved keeping a record of complaints; to monitor for resolution; whose responsibility it is to respond; and whether a case is closed or still being processed… An employee had an issue concerning a particular manager in Minera Exar and he approached his community leader rather than company management. The community leader then talked with management, which took corrective action and resolved the problem [male company employees]…

Community-company

The term "conflict" did not come up in any of the interviews with community members, and comments by external actors and by a member of one of the six communities at a CSR seminar suggest that community-company conflicts are rare and that, if something is not right, the communities will take a dialogue approach to resolve the issue:

> …We have never had a complaint about Minera Exar [provincial authority]…Now the company is a new neighbour that is incorporated and we open the doors to it, but without forgetting that we will be the first overseers of the agreed-upon precepts and if we see some errors we will rely on the dialogue that characterizes us to ask for explanations and corresponding rectifications [Speech by a female member of one of the six communities]…

4.5.4.4 Goal Compatibility

Within company

As this study focused on social responsibility, many employee comments relate to this topic. While the company is young and all employees were new and culturally different (about half of them came from other regions of Argentina), their view of the issues and their importance aligned remarkably well with stated company objectives.

...We need policies and concrete actions in CSR to meet our goals. Although sometimes there is no time, we have to prepare ourselves to work on what needs to be improved, to be ready for what is coming...The more I know, the more I can compare it to what we are doing to improve [with respect to e3 Plus]. The consultants also have to know e3Plus...We talk about it [e3 Plus], but we do not have time to get into the details [female employees]... The CEO is the first who needs to commit to CSR, followed by the managers who have to know what has to be accomplished in their area... This issue is becoming more critical and we need more CSR training. We need to prepare ourselves for new responsibilities and roles, and for the arrival of more people... Now we are like a child growing up with the values of the company. Seeing the good example of our superiors, we know what to do when we grow and continue as a larger company...CSR has a significant impact on my day-to-day work...Although we do not have a course yet, I want to learn about it and apply it. CSR is becoming more and more important to the managers...I know from informal conversations that we need to have good relations with the communities to avoid problems and obtain the social license to operate [male employees]...For the CEO CSR is as important as the rest of the business. Personnel support the CSR policy well...Our major challenges: become a productive mine and achieve our objectives. Continue working on the social as much as we can. Create a source of employment and leave a future for the people. We have done much in two years [male management employee]...They [the company] have moral force. They do the work to do what is correct – manipulating is bad...Safety and environment are the major challenges now and into the future [these are company priorities]...My major challenge is to be able to complete the objectives of the company [male employees]...

Community-company

A community member summarized the communities' understanding of their and the company's goals as follows:

> ...At the beginning the people of the six communities of the Department of Susques involved in the lithium project after the company had provided us with information on the work they thought of doing, after having visited the site where the work would be done, and after having asked the authorities in Jujuy [the provincial capital San Salvador de Jujuy] what they thought of it, worried about the impacts, we decided to ask and listen. In this way, we formed ourselves an opinion of what the work is and of what role we need to assume in the development of this mining activity in our territory...Now we bet on this activity, generating change, and if we are mistaken, we will see the error we made to correct it. We neither want nor will allow that others "key in" our future. To the contrary, we invite them to be participants in the sustainable development that comes hand in hand with respect [female community member]...

These words are a textbook example of how symbolic interactionism works. The project arrived and the communities gave it the joint meaning of "something to be looked at" (some communities were initially opposed and some were not). They debated the issue (interaction with their peers), drew on their reference communities (the company, the authorities in San Salvador de Jujuy and their own communities) and arrived at a more definitive meaning. It took between one and a half and two years of intense debate for the meanings of the communities to converge. They then decided to "take the jump" and make changes along the way as needed. While the processes to achieve convergence of meanings within the company and within the communities were quite different, the time taken was of the same order of magnitude.

4.5.4.5 Balance of Power

Within company

The first two comments quoted below suggest that the balance of power clearly slants towards management, whereas the subsequent quotes indicate that there still is considerable freedom for individual employees to act.

> …Structural relations regulate daily activities… Limits are established appropriately and the company is well organized – it delegates to the various areas of work and has good practices [male management employee]… I am proud to work for a company that wants to do things well, gives people the opportunity to do things well, and has confidence in us… There is respect within the company… Higher level employees trust lower level employees and let them work freely [male employees]…

The work of Bowen, Newenham-Kahindi, and Herremans (2010), Kemp et al. (2012), Kemp and Owen (2013) implicitly suggests that balance of power within companies may affect the effectiveness of their CSR practices, which supports the relevance of the above observation.

Community-company

The indigenous lands on which the project is located have been "measured and handed over" to the communities under article 75 in. 17 of the Argentine constitution. This means that they have full title and the right to decide what can happen on their lands. Without their explicit permission, the company cannot carry out any activities, a fact of which the company is well aware:

> …We establish our relationship with the communities and their understanding and permission allows us to go ahead with our projects – the so called social license to operate… Dedicating resources to community engagement and development is an investment. Without it, the communities will not allow the company to operate [male management employee]…

The communities own the land and have the power to shut the project down. The company brings resources and skill sets that the communities lack and from which they hope to learn and benefit. Both parties are fully aware of this and act accordingly. To be clear, "power" in this context does not link to "struggle", but rather to who brings what and who can most influence what processes, for good or for bad. The agreements signed with the communities appeared to provide a clear description of responsibilities and expectations and comments by community members (quoted earlier) indicated their satisfaction with the company meeting its end of the deal. The company in turn expressed its desire for the communities to be together with them in this venture and appeared to be satisfied with how things were playing out.

4.5.4.6   Focus and Frequency

While I was not able to review the agreements between the company and the communities, the community member comments cited above and many of the other community interviewee comments quoted earlier indicate that both parties have negotiated and are fully aware of their respective roles and responsibilities. The company is always present at the monthly general assemblies of the communities and the parties call additional meetings when issues arise. As I discussed in Chap. 3, visibility; sincerity and personalization; showing face; and establishing routines lead to the development of relationship patterns and new social structures surrounding mineral exploration projects, and to determining the course of social events and the perceptions of present and future benefits and harms. The Cauchari-Olaroz project appeared to meet all these criteria, which boded well for the future. However, external events sometimes interfere as they did in this case: financing became difficult and there was disagreement with the provincial government about the conditions under which the newly established provincial mining enterprise could become a partner in the project. Because of these difficulties, the project ceased operations for one and a half years and activities resumed in the fall of 2014. The company-community relationship survived these setbacks. The interactionist processes dealt with the changing circumstances through the relationship and interaction patterns that the actors had established and that allowed all parties to adjust their meanings to the new circumstances without too much divergence.

4.5.4.7   Stability

Within company

There were no specific interviewee comments linked to the predictability of relationships within the company. However, many of the comments quoted earlier paint a picture of generally agreeable relationships, and it is reasonable to assume that "agreeableness" correlates positively with predictability.

Community-company

The structures and practices mentioned under "focus and frequency" established a platform for predictability. While there may have been bumps, everyone was confident to find a way through these using the established patterns and processes.

4.5.4.8   Productivity

Within company

Earlier interviewee comments indicated that the target results of the company were to establish good relations with the communities, contribute to their well-being, obtain their permission to work and to begin producing lithium. They achieved the first three target results, as indicated by community interviewee comments. They did not yet achieve the latter, for reason beyond the scope of the present study.

Community-company

The community-company relationship met the criteria outlined by the member of the community of Puesto Sey that were mentioned earlier: the company and the communities acted as good neighbours, and if there were any issues (no substantial ones were raised in the interviews), they must have been resolved through dialogue. From the company perspective, the relationships allowed the company to do its work.

4.5.5 Reference Communities of the Company and Its Employees

The most relevant way of categorizing subsets within the company was by "employee origin" and by "rank", leading to the subsets: "local employees"; "non-local employees"; and "management". Current reference communities of management likely included company head office, relevant professional organizations, peers in mineral exploration and mining. At the time of the study, the "company" reference communities were probably nearly identical to the "management" reference communities.

As mentioned earlier, this may change with time as, by nature, relationships are not static. In adopting the Equator Principles (2013) and later e3 Plus for its social responsibility approach, company management was measuring itself against the norms of professional organizations and peers.

Local employees' reference communities likely consisted of the local community members and leaders; the general assembly; religious and spiritual leaders; and family networks, as is evident from an employee's decision to approach the community leader for help with solving a problem at work, and employee concerns about the company not offering to the PachaMama (the earth/time mother goddess). At the same time, several local employees were well aware of and influenced by company norms and values, as part of the dynamic change process.

The subset of non-local employees (mostly from other provinces of Argentina) probably did not have a single group of reference communities: each related to his or her reference community "of origin". With time and continuing interaction, "the company" may become more and more included in the subsets' reference communities.

4.5.6 Reference Communities of Community Members

Real and imaginary members of the reference communities of community members likely included general assemblies, other communities in the zone, tribal organizations, umbrella indigenous organizations, spiritual leaders, oral traditions and possibly certain NGOs. This aspect needs more study. A community member represented the six communities at the social responsibility seminar described in the next section. She skilfully melded the ideas and concerns alive in her communities with the norms on indigenous issues being articulated and promoted through organizations such as the International Labour Organization, the United Nations, the International Finance Corporation and others to explain why the communities had decided to support the company and what safeguards they were putting in place. However, she did not make any direct reference to these norms: they had become part of the broader global indigenous reference community. Aspects of this reference community overlap with those of the professional and peer reference communities of the company (both probably drawing on the same sources), which definitely facilitated reaching an agreement.

4.5.7 Company Social Responsibility Approach

This section describes the strategic approach of the company and the communities.

In September 2012, both the CEO of Minera Exar and a member of the community of Puesto Sey (one of the communities in the area of influence of the Cauchari-Olaroz project) participated in a seminar on social responsibility during Argentina Mining (held in Salta). The text below draws on their observations and on interviewee comments.

For the company, social responsibility was a business vision that integrated the management of the company with respect for the values and ethical principles of the employees, the community and the environment. Preparations included surveys, a socio-environmental baseline study, hiring of specialists with experience in the subject matter and in the area, and involving senior personnel in the work area.

In the design of its social responsibility approach, the company drew on the Equator Principles and e3 Plus (Pérez, 2012). The plan included the following components: local purchasing, local employment where possible, support for local initiatives, communications and consultation. Lessons learned: the most affected communities need to benefit most, it is better to be open and clear than to be secretive, and many small "public work projects" have more impact than one big one. Big "public works projects" would compete with the state or other social actors and could cause more problems than benefits and cooperative undertakings are more effective than gifts. It is important to implement a social responsibility plan *before* problems occur, not *when* they occur (Pérez, 2012).

From the communities' perspective, the change processes had strong links to the right of the communities to free, prior and obligatory consultation about the implementation of mining activities on their lands. The change process also enabled them to determine if the project would affect or damage their interests through its environmental, social, economic and cultural impacts. They were well aware of the changes that are occurring in the mineral exploration and mining industry and of the increased expectations placed on companies:

> …The situation has changed and nowadays there are companies that are responsible and that recognize that it is a duty to treat communities with respect and solidarity, and that the use of good practices is the key to sustainable development. Often, the adverse impacts of mining in the past have been what weighed most on the memory of the people, for which reason indigenous populations have set out rules of the game that are more strict. They are now ensuring that their right to respect for their territories, culture and society prevails…It is now common practice that companies request permission to explore; communities visit the site during project development; companies participate in general assemblies and cultural events; the companies develop local suppliers and provide capacity building to employees and the community; companies treat their employees well in all respects… [female community member presentation at a CSR seminar]…

The six aboriginal communities expressed the following expectations:

> …Ongoing participation of the state and companies in the development of the region jointly with the communities; priority employment; joint community-company-state environmental protection; existence of good practices to protect the health and safety of the workers and the local population; integral education and training plan; family strengthening; respect for local laws and customs; easement contract compliance; a social impact study aimed at the long term; dialogue and fluid communication between the community and the company; push the government to return part of the taxes and royalties to the region; plan for direct infrastructure investment; and rational exploitation of natural resources [female community member presentation at a CSR seminar]…

The communities had engaged in an intense process to arrive at their decision:

> …We know that we are just at the beginning, but we have already been working together, state, six communities and company (some more than others) but I understand that our role is very important. That we have to develop it with responsibility, communicating all information we receive from the state and from the companies, while at the same time make it clear what are the concerns of the people from the communities. This has been a long process in which we all have participated, through which we got to know each other, we have debated, we have demanded, but always with respect and from the place that each of us occupies…We are not beholden to the companies. Yes, we are engaged with social responsibility, with the development of our communities. Now the company is a new neighbour that joins us and we open the door to them, but without forgetting that we will be the first to supervise the agreed-upon precepts and if we see any errors we will use the dialogue that characterizes us to ask for the corresponding explanations and rectifications…Now we are betting on this activity, making the change and if we are mistaken, we will look at it to correct it. We do not want nor will we permit that others determine our future. On the contrary, we will invite them to be participants in the sustainable development that comes with respect [female community member presentation at a CSR seminar]…

As was seen earlier, interviewee comments generally agreed with the sentiments expressed above and they felt positive about the company's initiatives and their involvement, with some exceptions and specific concerns.

4.5.8   Perceived Present and Future Benefits and Harms

Benefits to the community included the income from compensation for the use of their land (under "convenios de servidumbre"). Other benefits mentioned included: support of the NGO SUYAI in various campaigns in the battle against cancer, collaborating with the transportation of women to the location of the truck sent by the Ministry of Public Health; capacity building; employment; road development and cleaning; various types of support to educational, religious, sports and municipal institutions. While the company was quite prepared to help with issues such as garbage disposal and sewer systems, they would only do so if the communities took the initiative.

Benefits to the company include approval of the Environmental Impact Report within eight months; strong community support for construction of the mine; signed agreements with each community, including an integral development plan; united support from local communities against numerous attempts to attack the project by outside groups (Pérez, 2012).

Some community concerns remained:

> …For now they respect us, we do not know about later on…The respect for the local people has to do with company personnel. Let's hope that the same level of respect is maintained if the personnel changes…" Hopes for the future include "…That the company maintain its participation, that the workers have better relations and that our culture be respected, that they participate in the PachaMama, and that the meetings be a little more extensive so that people better understand their technicians and specialists …sometimes we see barriers because we have other ways of saying things – simple words – drawings are needed [male community members] …We would like them to address our concerns with independent external help [female community member] …

In addition, some of the communities outside the area of influence of the company complained about being left out.

4.5.9   Status in September 2019

The Cauchari-Olaroz project anticipates starting production of lithium carbonate by the end of 2020 and projects a mine life of 40 years (Lithium Americas, 2019). Its web site does not mention community relations.

4.6   Lindero

4.6.1   Field Study Information

The study took place from 11 to 15 October 2011. Interviews and conversations were held in Salta (the provincial capital), San Antonio de los Cobres, Tolar Grande, the Arita exploration camp and Cavi (an oasis at 20 km from the Arita camp). Because of time limitations, some of the interviews and conversations were conducted in a group format (e.g. the interview with about 10 staff in the Arita camp and the conversation with members of the Chamber of Mines). While conversations related to the topic of study, they were free flowing and did not follow the interview guide. I interviewed community members, municipal and provincial authorities, a politician, a judge of the peace and company staff. Of the 36 people interviewed, 17 were company personnel, 8 were female and 28 male. Respondent distribution skews towards Mansfield Minera employees and professionals, and a number of interviews and conversations took place in Salta. The logistics of travel between the various locations and the absence of accommodation in Tolar Grande played an important part in this respect. However, there were no significant contradictions between comments across respondent categories and a coherent picture of relationships emerged.

Mansfield Minera had been active in Salta since 1994, and in 1999, they discovered the Lindero prospect that reached the prefeasibility stage in 2010. The company had a working capital of C\$4.4 million (Thomas, Melnyk, Nimsic, Khera, & Kappes, 2010). The Lindero project is located in the puna, in the Department of Los Andes, province of Salta, Argentina. It is accessible by road (paved part of the way) from Salta, through San Antonio de los Cobres. The driving distance is about 420 km, which takes some 8 h. The project is located at 75 km south-west of Tolar Grande, the town closest to the project. It lies at an altitude of approximately 4500 m above sea level (Ausenco Vector, 2010). The Arita field camp is located on the edge of Salar Grande.

4.6.2   Project Description

The principals and the director have worked with the company since the beginning. The company managed finances and resources very cautiously over the life of the project, as is evident from the somewhat unusual length of time it took to develop the project (more than 16 years as compared to an average of four to eight years for most projects). Sufficient patient shareholders stayed on board to continue the project. The cautious approach paid off in terms of establishing stable community relations. Vancouver head office fully trusted the local office and gave it great freedom of action while at the same time setting clear and sufficiently broad boundaries. The local director was also the president of the Salta Chamber of Mines, a function that, with approval of head office, took 70% of his time. This role connected him very well to many sectors of Salta society, and he had thorough background knowledge of much that went on in the Province. The company had an office in the city of Salta that handled purchases, administration, government relations, core shack and logging of core. The Arita field camp was located close to the project site and consisted of a number of portable buildings that contained an office, a meteorological station, sleeping quarters, a diesel electricity generator, showers and a dining room.

4.6.3   Context

The total population of the province of Salta was 1,215,207 in 2010. The Department of Los Andes in which the project is located had a population of 6126. The average population density of the province as a whole was about 7 inhabitants per square kilometre, while for the Department of Los Andes it was only 0.2 (Instituto Nacional de Estadística y Censos, 2010a). There are two municipalities in the Department of Los Andes: San Antonio de los Cobres and Tolar Grande with 5482 and 148 inhabitants (2001 census), respectively (Romero & González, 2008). San Antonio de los Cobres is important from a company and CSR perspective because it is the administrative centre of the department. While it was not possible to access recent census data through the Internet, the population of Tolar Grande at the time of this study was around 168, indicating stabilization since 2001.

The percentage of people living below the poverty line in Salta was 13.4% in 2001, compared to a national average of 9.4%, and 4% were living below the extreme poverty line, as compared to 4.8% nationally (Instituto Nacional de Estadística y Censos, 2010a; Instituto Nacional de Estadística y Censos (INDEC), 2001b). The employment data show that a very high proportion of Tolar Grande's labour force was working in the public sector. The borax mines in the province have had a long (70 years), stable relationship with the communities and are part of them.

All participants in a meeting with the Salta Chamber of Mines recognized the importance of coordinating activities between companies. The Chamber organized annual events in San Antonio de los Cobres in which each company showed what it was working on and talked about its plans. Local employees made the presentations. The events were combined with capacity building in the afternoon, so far mostly for municipal employees, emergency personnel and the like (e.g. ATV operation, defensive driving). While a number of provinces had prohibited open pit mining, Salta had not done so. The local culture is strongly religious (Catholic and indigenous beliefs, thoroughly mixed), family-oriented and machismo reigns. Companies need to hire local people to do well in CSR—culture is extremely important.

4.6.4 Observed Patterns and Characteristics of Relationships

As a relatively large number of employees participated in the interviews, it was possible to compare relations within the company with those between the company and the community. The sections below summarize the comparison.

4.6.4.1 Trust and Respect

Within company

The interviewee comments cited below imply a high degree of mutual trust and respect:

> ...the boundaries are clear and sufficiently broad; the company is thorough and patient ...I trust their work completely and send them out to prospect on their own – they played an important role in the discovery of the Lindero and Arizaro prospects [male manager] ... We like working for Mansfield because we know what is expected and we are left free to decide how we want to go about achieving our goals [male employees]...There is nothing new about CSR – we have been taking this approach for the past 14 years..Management always made it perfectly clear that all people should be respected and treated well, whoever they are...We act as the eyes and ears of the company and when we see or hear something that falls outside our [wide] range of responsibility we advise the director who then takes action at a higher level [male employees]...Duties have to be shared within logical limits...Hierarchy exists but it is built on context, experience, agreement and understanding [male managers]...I had been offered positions in other companies but I set my mind on joining Mansfield Mineral because of its good reputation [female employee]...Things always get talked over and there is a good dialogue 90% of the time...The company is very flexible in solving problems and is going very easy on employees [male employees]…. Life in the camp is like a family. It is important to provide good service and treat everyone with respect [female-male employee team]...

Community-company

The comments cited below indicate that there was trust and respect between the community and the company. However, the number of community members interviewed was small, and although company employees who are also community members made many of the comments, this assumption is questionable. Still, I observed that, when driving through Tolar Grande and through San Antonio de los Cobres, community members frequently flagged down the company truck to greet its occupants and discuss matters of interest or concern. This lends support to my assumption.

> ...As Mansfield has been around for a long time, community expectations are realistic and we can say "no" to certain requests for assistance... We know the people of the puna and they want companies to behave well ... There have never really been problems with the community... We are part of the community and talking usually resolves any problems... On a number of occasions, the community itself resolves problems, for example, when some people said certain things about the company and others did not agree, community members resolved the matter amongst themselves and it was settled... I [Mansfield director] often receive calls about community problems even before the intendente [elected leader of Tolar Grande] does [company manager]... The relationship between the companies and the

municipality is good and there is an open dialogue … The community is well aware that
the opening of a mine will cause enormous changes, and they are preparing through their
municipal plan … The municipality and the companies regularly meet to discuss and develop
civil defense issues, plans and activities … The regular civil defense meetings between the
municipality and companies are useful [male members of the local authority]…We would
like Mansfield Minera to work with us, but the mayor is opposed [female member of a
community group in San Antonio de los Cobres]…

4.6.4.2   Communication and Mutual Understanding

Within company

Employee comments suggested that there is good communication and mutual
understanding within the company:

> …Management decides on most CSR issues and actions. Everyone is aware of these decisions
> and they consider them in logistics, purchasing and other company activities. There is open
> communication about all this [male manager]… Suggestions and issues are discussed as a
> group… We work as a team and rotate tasks on the basis of mutual understanding [male
> employees]…

Community-company

A number of the comments cited under previous indicators also have a bearing on
communication. Both those and the additional relevant comments cited below suggest
that there was good communication and mutual understanding between the company
and the community:

> …All actors are important. We know everyone and everyone is important… Tolar Grande
> is fairly united. In spite of this, it still takes much effort to get the people together… There
> is an open dialogue [between the community and the companies]… Most issues are set-
> tled in a "natural" way… The village fiestas (religious, indigenous, mother's day, and cul-
> tural occasions) are very helpful in this respect – many things are talked about [municipal
> authorities]…

4.6.4.3   Conflict Resolution

There are no formal conflict resolution mechanisms either within the company or
between the company and the communities, but interviewee comments cited earlier imply that there was little conflict and that existing relationship patterns presented ample avenues for addressing conflict situations either within the company or
between the community and the company:

> …sometimes frictions do occur like in any family or community. We are part of the commu-
> nity and talking usually resolves any problems… Community members resolve the matter
> [related to the company] amongst themselves and it is settled… there is an open dialogue…
> most issues get settled in a "natural" way [male local company employees]… Decisions

are usually made quickly. The village fiestas (religious, indigenous, día de la madre, cultural occasions) are very helpful in this respect – many things are talked about [male local authority]... The company is very flexible in solving problems and is going very easy on employees... Problems, if they occur, are usually small. It is very important to maintain good relations... when problems crop up; we usually talk them over as a group. The group leader decides how to approach the situation based on the discussion. There have not been any significant problems that we can remember [male local employees] ...

4.6.4.4   Goal Compatibility

Obviously, the company wants to establish a mine. It also wants to contribute to the community:

> ...It is part of our duties to contribute to the community [male local employee]... In December, there will be a new intendente in San Antonio de los Cobres with whom Mansfield has an excellent relationship and we hope that we will be able to contribute more to the community... We are a company with a real interest in the prospect and its characteristics and in the region and the communities... When the Environmental Impact Assessment gets approved, Mansfield Minera faces a tough choice: do we sell the prospect to a mining company (thereby potentially putting our obligations to our employees and to the communities at risk) or do we convert into an operating company (which involves a huge risk as we do not have any operating experience) [male company manager]...

The community is also preparing for the opening of a mine:

> ...We are well aware that the opening of a mine will cause enormous changes, and we are preparing through our municipal plan [male local authority]...

4.6.4.5   Balance of Power

Within the company

The employee and management comments cited earlier suggest that, while management decides on major issues, it is close to its employees, seeks and uses its input and leaves ample room for employee decision-making. The balance is somewhat tilted towards management, but of all the case studies, it is probably closest to equilibrium.

Company-community

The community has some regulatory power over the company:

> ...The municipality carries out safety, environmental and food quality checks on companies [female local authority employee]...

It is laying out its own plans for a mining future (see earlier quote) and did not appear to have the same power over its lands, as did the communities surrounding the Cauchari-Olaroz project but I was not made aware of the details. Rather than looking at the community and the company as entirely separate entities, it may be more appropriate in this case to say that the company is a community member that has more resources, knowledge and expertise than do many other community members and that does not "throw around its weight".

4.6.4.6   Focus and Frequency

Within company

Comments cited earlier suggest that everyone in the company was aware of expectations and company goals. While I did not receive information on the specifics about meeting frequency, there appeared to be continuous interaction on significant issues.

Community-company

> …Meetings with the community are held 3–4 times per year or when the need arises. In addition, the company participates in all religious and cultural events, and many employees are community members [male company personnel]…

Earlier quotes mentioned additional informal occasions that provided an opportunity to discuss matters, such as festivities and cultural events. The company also took part in the monthly community assemblies and regularly met with the local cacique.

4.6.4.7   Stability

Within company

The company has been around for some 20 years, and its practices have settled into predictable patterns. Several of its employees represent successive generations of one family.

Community-company

While there were no specific comments by members of the community that related to this matter, many of the comments cited earlier suggest that the relationship was stable.

4.6.4.8 Productivity

The target result of both the company and the community was to see a mine in operation and to maintain good relations. The company's management style and the affective skills of its management and employees led to excellent relations with Tolar Grande. It also led to significant mineral discoveries and impending construction of a mine (equipment had been ordered). The company's target of being able to contribute to San Antonio de los Cobres came a step closer with the appointment of a new intendente who immediately hired Mansfield Minera's CSR manager to help him with social issues (personal communication, Facundo Huidobro). Mansfield Minera frequently communicated with the legislative, juridical and executive branches of government and with the Catholic Church. The company played an active role in the Salta Chamber of Mines. It had been a driver for coordinating activities between companies, and the Chamber had started a CSR working group for this purpose. It viewed social responsibility as something shared by all actors, a duty for all. As a result, indigenous organizations in areas where more than one company is active have signed an agreement with the Salta Chamber of Mines rather than with individual companies (Ministerio de Ambiente y Producción Sustentable, 2012). The mineral exploration companies active in the area and the municipality of Tolar Grande held regular meetings where they discussed emergency and safety and security measures and agreed on related protocols. For example, they produced a map of the area that showed the location of roads, mining camps, areas covered by cellular phone towers, shelters, nature reserves, etc. On more than one occasion, Mansfield Minera and other exploration companies had to salvage tourists that became stranded "in the middle of nowhere". Mansfield Minera had become part of the community of Tolar Grande, partly because the company had been there for a long time and partly because it had paid much attention to building relationships. In terms of relationships:

> …Our most important relationships are those with the intendente [elected leader] of Tolar Grande and the intendente of San Antonio de los Cobres. Tolar Grande is located at 75 km from the project and San Antonio de los Cobres at 175 km. Even though the latter is far away, the relationship is important because it is the administrative centre of the Department. Its intendente has been negative towards company involvement with the community and has asked all companies to stay out of town. He also subjected them to a levy…Other important relationships are those with the cacique [leader of the indigenous people] in Tolar Grande and school personnel [company manager]…

The elected cacique coordinated all activities related to indigenous issues in Tolar Grande (cultural, inclusion, etc.). These observations lend credence to the assumption that Mansfield Minera's internal relationship dynamics greatly influenced the dynamics of its relationships with the community. They also led to a very low risk-of-conflict measure on all relationship indicators, as shown in Table 5.1 in Chap. 5 that will discuss the interactionist aspects of the Lindero project in more detail.

4.6.5   *Company Social Responsibility Approach*

Mansfield Minera's management was of the opinion that:

> …Rules-bound companies that have a "tight" management style are not as interested in or as good at CSR as companies that are run in a more flexible way [male company manager]…

As will be clear from much of the foregoing, Mansfield Minera's management style was anything but tight and rules-bound: the company entrusted its employees with demanding tasks and gave them much flexibility in the way they chose to complete these tasks. With respect to CSR, several employees specifically mentioned that:

> …It is part of our duties to contribute to the community [male company employees]…

As I mentioned earlier, the company approached CSR in an informal way, but everyone knew about it and they discussed issues and suggestions as a group. The company had developed its corporate capacity, complied with legal requirements, and ensured that its contractors met standards. It also had interacted intensively with governments and the community:

> …We maintain close links with all branches of government: legislative, juridical, and executive and also with the church. As a result of this outreach, we are often is called by members of the various branches of government who ask for information [male company manager]…

It did not appear to have had a need to draw on civil society expertise and resources to meet its CSR objectives. The company's approach to social responsibility grew organically from its beginnings at a time when Corporate Social Responsibility was an almost unknown concept in mineral exploration and mining. Only a few years ago did the company formally record a set of practical social responsibility guidelines for day-to-day actions, and it did not see a need to formalize the value system that is implicit in the way it goes about internal and external business (in contrast to companies such as Kinross) (Huidobro, 2011). Mansfield Minera paid attention to diplomacy:

> …Diplomacy is of the utmost importance… Our employees are our ambassadors in the community and it is important that their behaviour be beyond approach [male company manager]…

This mostly worked well, but as the company had no "jurisdiction" over employee behaviour outside work, certain situations posed a challenge in this respect:

> …We are not sure what to do about the worrisome behaviour of one employee outside work [paraphrased somewhat for confidentiality reasons] [male company manager]…

The company had identified health, education and infrastructure as priorities for community support to help manage the enormous amounts of requests it was receiving (Huidobro, 2011).

4.6.6 Perceived Present and Future Benefits and Harms

Examples of benefits the community derived from Mansfield Minera's presence include:

> …All staff takes part in the vaccination of the herd of llamas, goats and sheep at the oasis close to the Arita camp. The company provides transportation, vaccine and organization. The province provides the veterinarian…It has always been company policy to transport community members who want to travel to Salta or San Antonio de los Cobres…The roads constructed by the exploration companies [including Mansfield Minera] are making a huge difference…The company participates in all religious and cultural events [male company personnel]…

Additional benefits include replacement of the drainage gallery of the drinking water intake jointly with the municipality and other companies after the case study had been completed (personal communication, Facundo Huidobro). Mansfield Minera opened the Arita camp cafeteria to tourists and contributed to road construction and signage. It supported religious and indigenous events and festivities, and provided equipment and supplies for the medical post and the schools. In addition, it participated in general eyesight testing and follow-up in cooperation with other companies, the hospital of San Antonio de los Cobres and the Salta mining Secretariat. Capacity building in areas such as masonry, carpentry, electricity, blasting and heavy equipment maintenance, in cooperation with foundations and other exploration companies, was also part of its CSR programme (Huidobro, 2011).

Main community concerns in the area centred on issues like:

> …My main preoccupation is the depopulation of the town because of lack of opportunities [the Chamber of Mines was discussing this problem and looking at things such as jobs and scholarship schemes tied to the town]… I am very concerned about the water supply to Tolar Grande… My main preoccupations are water (the present source is running low) and capacity building – the community's capacities are only so-so… What will happen when the ore body is exhausted and the company will leave? [male local authority]…

As mentioned in the previous paragraph, the exploration companies active in the area got together and solved the water supply problem after I conducted this case study.

4.6.7 Status in September 2019

Fortuna Silver Mines acquired Goldrock and its Lindero project in 2016. Construction of the mine began in 2017, and the company plans the first doré pour for the first quarter of 2020 (Fortuna Silver Mines, 2019). The company's web site is silent on community relations.

References

Administrador. (2011). Prefecto Carrasco plantea declarar a los páramos del país libres de minería. Retrieved October 21, 2013 from http://cadenaradialvision.com/index.php?option=com_content&view=article&id=1414:prefecto-carrasco-plantea-declarar-a-los-paramos-del-pais-libre-de-mineria&catid=1:locales&Itemid=2.

Agreda Orellana, W. G. (2011). *Inventario de las necesidades básicas insatisfechas y conflictividad social de las comunidades del área de influencia del Proyecto Estratégico Nacional Zarza, Provincia de Zamora Chinchipe, Cantón Yanztaza, Parroquia Los Encuentros* (Magíster en Gestión y Desarrollo Social, Universidad Técnica Particular de Loja), pp. 1–62.

Alonso, R. N. (2010) Jujuy: Los Jesuitas y la minería. Retrieved from http://www.casemiargentina.org/index.php/noticias/1035-jujuy-los-jesuitas-y-la-mineria.

Arias, L. (2013). El aporte económico de la minera Kinross se siente en Los Encuentros. Retrieved July 15, 2013 from http://www.elcomercio.com/negocios/economico-minera-Kinross-Ecuador_0_944305639.html.

Asamblea Constituyente, República del Ecuador. (2008). Mandato Constituyente No. 6. Retrieved February 6, 2015 from http://www.superley.ec/pdf/mandatos/06.pdf.

Atlas.ti. (2012). A world of data in your hand, atlas.ti 7. Retrieved January 20, 2013 from http://www.atlasti.com/index.html.

Ausenco Vector. (2010). *Informe de Impacto Ambiental Proyecto Lindero, Mansfield Minera S.A., Capítulo II: Descripción del Ambiente.* (Client report). Buenos Aires: Ausenco Vector.

Banco Central del Ecuador. (2011). Estadísticas económicas. Retrieved June 25, 2013 from http://www.bce.fin.ec/contenido.php?CNT=ARB0000003.

Bidaseca, K. (2011). World heritage listing policies, local knowledge and indigenous peoples in Argentina's Quebrada de Humahuaca. Retrieved August 18, 2014 from http://www.inter-disciplinary.net/wp-content/uploads/2011/08/bidasecampaper.pdf.

Boon, J. (2015). *Corporate social responsibility, relationships and the course of events in mineral exploration—An exploratory study* (Unpublished Ph.D. thesis). Carleton University, Ottawa. https://curve.carleton.ca/system/files/etd/6c6598d4-c436-409e-9ba1-40dea2d37d2c/etd_pdf/7b39ca613ff7e7e2df52ed82580e3974/boon-corporatesocialresponsibilityrelationships.pdf.

Borghini, N. (2010). Tenencia precaria de la tierra y políticas públicas en Jujuy, Argentina. Un análisis de los vínculos entre provincia, nación y pueblos originarios. Retrieved August 18, 2014 from http://www.up.edu.pe/revista_apuntes/SiteAssets/Natalia%20Borghini%20Apuntes%2067.pdf.

Bowen, F., Newenham-Kahindi, A., & Herremans, I. (2010). When suits meet roots: The antecedents and consequences of community engagement strategy. *Journal of Business Ethics, 95*(2), 297–318.

Centro Peruano de Estudios Sociales (CEPES). (2010). *El Tratamiento Legal de las Comunidades Campesinas.* Retrieved December 11, 2010 from http://www.cepes.org.pe/legisla/legisla.htm.

Claverías Huerse, R., & Alfaro Moreno, J. (2010). *Mapa de actores y desarrollo territorial en la Cuenca Lurín.* Lima, Perú: Centro de Investigación, Educación y Desarrollo.

Davis, R., & Franks, D. M. (2011). The costs of conflict with local communities in the extractive industry. In *First International Seminar on Social Responsibility in Mining, October 19–21, Chapter 6*, Santiago, Chile, pp. 1–17.

Deaux, K., & Martin, D. (2003). Interpersonal networks and social categories: Specifying levels of context in identity processes. *Social Psychology Quarterly, 66*(2), 101–117.

Defensa de Territorios. (2011). Ecuador: Tarqui y Victoria del Portete dijeron "No a la minería". Retrieved October 21, 2013 from http://www.diplomaciaindigena.org/2011/10/ecuador-tarqui-y-victoria-del-portete-dijeron-%E2%80%9Cno-a-la-mineria%E2%80%9D/.

Del Castillo, L. (2006). Property rights in peasant communities in Peru. Retrieved August 5, 2014 from https://www.mpl.ird.fr/colloque_foncier/Communications/PDF/Del%20Castillo.pdf.

Ekos. (2012). El consejo de minería responsable presenta el código de ética. Retrieved February 9, 2015 from http://www.ekosnegocios.com/NEGOCIOS/verArticuloContenido.aspx?idArt=1109.

El Tiempo. (2012). Paul Carrasco anuncia su precandidatura a la presidencia. Retrieved October 21, 2013 from http://www.eltiempo.com.ec/noticias-cuenca/99757-paa-l-carrasco-anuncia-su-precandidatura-a-la-presidencia/.

El Universo. (2013). Carlos Pérez Guartambel: 'Radicalizar resistencia, único camino para la supervivencia'. Retrieved January 5, 2016 from http://www.eluniverso.com/2013/04/23/1/1355/radicalizar-resistencia-unico-camino-supervivencia.html.

Equator Principles. (2013). Equator principles—environmental and social risk management for project finance—the Equator Principles III. Retrieved July 31, 2014 from http://www.equator-principles.com/index.php/ep3.

Fernández J. P. (2019). Proyecto Palma tiene el Potencial de un Cerro Lindo. Retrieved September 14, 2019 from https://www.energiminas.com/proyecto-palma-tiene-el-potencial-de-un-cerro-lindo-o-un-colquisiri-volcan-compania-minera/.

Flickr. (2011). Sistema comunitario agua potable VP-T. Retrieved October 21, 2013 from http://www.flickr.com/photos/62081634@N05/6215588294/in/photostream/.

Fortuna Silver Mines. (2019). Lindero Project, Argentina. Retrieved September 14, 2019 from https://www.fortunasilver.com/mines-and-projects/development/lindero-project-argentina/.

Gerencia Regional de Planeamiento. (2006). Informe técnico nº -2006-GRL/GRPPAT/AT - Informe técnico de recategorización del Centro Poblado de Tierra Blanca. Retrieved August 4, 2014 from http://www.regionloreto.gob.pe/OATSIG/Tierra_Blanca.pdf.

Gobierno Parroquial de Los Encuentros. (2011). *Plan de Desarrollo Y Ordenamiento Territorial – "Resumen Ejecutivo"*. 2011: Gobierno Parroquial de Los Encuentros.

Guartambel, C. P. (2012). *Agua u oro - Kimsacocha, la resistencia por el agua*. Cuenca, Ecuador: Carlos Pérez Guartambel.

Henderson, R. D. (2009). *Fruta del Norte Project, Ecuador, NI 43-101 Technical Report*. Toronto: Kinross Gold Corporation.

Huidobro, F. (2011). *Historia de trabajo de responsabilidad social empresaria y relaciones comunitarias realizadas por Mansfield Minera S.A. desde 1994* (Internal report). Salta, Argentina: Mansfield Minera S.A.

Iamgold Technical Services. (2009). Quimsacocha gold project, Azuay Province, Ecuador, NI-43-101 Technical report. Retrieved October 21, 2013 from http://www.iamgold.com/files/operations/43-101%20Technical%20Report%20Quimsacocha,%20February%202009.pdf.

Ilustre Municipalidad de Cuenca. (2011). *Plan de Desarrollo y Ordenamiento Territorial del Cantón Cuenca. Tomo 2: Diagnóstico integrado y modelo de desarrollo estratégico y ordenamiento territorial*. Cuenca, Ecuador: Ilustre Municipalidad de Cuenca.

Instituto Nacional de Estadística y Censos. (2010a). Censo de población y vivienda 2010. Retrieved June 15, 2012 from http://redatam.inec.gob.ec/cgibin/RpWebEngine.exe/PortalAction.

Instituto Nacional de Estadística y Censos. (2010b). Necesidades básicas insatisfechas. Retrieved June 15, 2012 from http://redatam.inec.gob.ec/cgibin/RpWebEngine.exe/PortalAction?&MODE=MAIN&BASE=CPV2010&MAIN=WebServerMain.inl.

Instituto Nacional de Estadística y Censos (INDEC). (2001a). *Cuadro 4.13. provincia según departamento. hogares y población: Total y con necesidades básicas insatisfechas (NBI). Año 2001*.

Retrieved August 17, 2014 from http://www.indec.gov.ar/micro_sitios/webcenso/censo2001s2_
2/ampliada_index.asp?mode=38.

Instituto Nacional de Estadística y Censos (INDEC). (2001b). Censo nacional de población, hogares
y viviendas del año 2001. Retrieved August 18, 2014 from http://www.indec.gov.ar/micro_sitios/
webcenso/.

Instituto Nacional de Estadística y Censos (INDEC). (2010a). Cuadro P2-P. Provincia de Jujuy.
Población total por sexo e índice de masculinidad, según edad en años simples y grupos quin-
quenales de edad. Año 2010. Retrieved April 22, 2014 from http://www.censo2010.indec.gov.ar/
CuadrosDefinitivos/P2-P_Jujuy.pdf.

Instituto Nacional de Estadística y Censos (INDEC). (2010b). Distribución del ingreso - encuesta
anual de hogares urbanos - tercer trimestre de 2010. Retrieved August 17, 2014 from http://www.
indec.gov.ar/nuevaweb/cuadros/4/eahu_dist_ingreso_08_11.pdf.

Kemp, D., & Owen, J. R. (2013). Community relations and mining: Core to business but not "core
business". *Resources Policy, 38*, 523–531.

Kemp, D., Owen, J. R., & van de Graaff, S. (2012). Corporate social responsibility, mining and
"audit culture". *Journal of Cleaner Production, 24,* 1–10.

LibreRed. (2012). Ecuador: 30.000 personas marchan por el derecho al agua y en contra de Correa.
Retrieved October 21, 2013 from http://www.librered.net/?p=16198.

Lithium Americas. (2019). On the road to production. Retrieved 09/14, 2019, from https://www.
lithiumamericas.com/cauchari-olaroz/.

Lundin Gold. (2019a). Project overview. Retrieved September 14, 2019 from https://www.
lundingold.com/en/fruta-del-norte/project-overview/.

Lundin Gold. (2019b). Sustainability report. Retrieved September 14, 2019 from
https://www.lundingold.com>sites>assets>files>2017_sustainability_report_lug.

Martínez, R. G., Golovanevsky, L., & Medina, F. (2010). *Economía y empleo en Jujuy* (Documento
de Proyecto No. LC/W.344 LC/BUE/W.49, 7-8). Santiago de Chile: Comisión Económica para
América Latina y el Caribe (CEPAL).

Méndez Urgiles, X. S., & Patiño Enríquez, A. F. (2013). *Georreferenciación y análisis de la pobreza
urbana y rural de las provincias: Pichincha, Guayas y Azuay a través del método de necesidades
básicas insatisfechas, y contraste de sus resultados con una metodología multidimensional de
pobreza 2010* (Unpublished Economista). Universidad de Cuenca, Cuenca, Ecuador.

MICLA. (2012). Quimsacocha, Ecuador. Retrieved October 21, 2013 from http://micla.ca/conflicts/
quimsacocha/.

Ministerio de Ambiente y Producción Sustentable. (2012). Comunidades Collas de Los Andes
y la Cámara de la Minería firman acuerdo de cooperación. Retrieved February 7, 2015
from http://www.salta.gov.ar/prensa/noticias/comunidades-collas-de-los-andes-y-la-camara-de-
la-mineria-firmaron-acuerdo-de-cooperacion/17595.

No a la Minería. (2011). Plebiscito sobre minería se impulsa en Girón, Ecuador. Retrieved October 21, 2013 from http://www.noalamina.org/mineria-latinoamerica/mineria-ecuador/plebiscito-
sobre-mineria-se-impulsa-en-giron-ecuador-2.

Pérez, W. (2012). Aspectos financieros de la responsabilidad social empresaria. Retrieved March
22, 2014 from http://www.olami.org.ar/paginas/eventos/eventos_anteriores.asp.

Prospectors and Developers Association of Canada. (2010). e3PLUS—A framework for responsible
exploration. Retrieved December 28, 2010 from http://www.pdac.ca/e3plus/index.aspx.

Quispe López, R. (2009). Evolución de la pobreza en el Perú: 2009. Retrieved December 12, 2013
from http://censos.inei.gob.pe/DocumentosPublicos/Pobreza/2009/Exposicion_Jefe.pdf.

Romero, G. M., & González, R. I. (2008). *Anuario estadístico Provincia de Salta: Año 2007: Avance
2008.* Salta, Argentina: Dirección General de Estadísticas.

Sarudiansky, R. (2008). *Departamento de Susques* (Informe Interno, Centro de Estudios para la
Sustentabilidad). Buenos Aires: Universidad Nacional de San Martín.

Secretaría Nacional de Planificación y Desarrollo Ecuador. (2009). El Buen Vivir en la Constitución
del Ecuador. Retrieved January 20, 2013 from http://plan.senplades.gob.ec/3.3-el-buen-vivir-en-
la-constitucion-del-ecuador.

Thomas, D. G., Melnyk, J. C., Nimsic, T. L., Khera, V., & Kappes, D. W. (2010). Lindero project, Salta province, Argentina, NI 43-101 Technical Report. Retrieved August 18, 2014 from http://www.goldrockmines.com/i/pdf/reports/Lindero-TR-May-2010.pdf.

Unidad de Desarrollo Social. Perú: Programa no escolarizado de educación inicial (PRONOEI). Retrieved August 5, 2014 from http://www.oas.org/udse/wesiteold/peru.html.

Velásquez, T. A. (2012). The science of corporate social responsibility (CSR): Contamination and conflict in a mining project in the Southern Ecuadorian Andes. *Resources Policy, 37*(2), 233–240.

Volcan Compañía Minera S.A.A. (2010). Proyecto de exploración minera Palma, Declaración de Impacto Ambiental, Resumen Ejecutivo. Retrieved December 12, 2013 from http://www.scribd.com/doc/145182443/RE-1977429.

Volcan Compañía Minera S.A.A. (2012). Relaciones comunitarias y medio ambiente. Retrieved December 12, 2013 from http://www.volcan.com.pe/site/#.

Volcan Compañía Minera S.A.A. (2013). Junta obligatoria anual de accionistas, lima, 20 de Marzo de 2013. Retrieved December 12, 2013 from http://www.slideshare.net/ernestolinares9/volcan-compaa-minera-2012.

Volcan Compañía Minera. (2019). Retrieved September 14, 2019 from https://www.volcan.com.pe/operaciones/mineria/exploraciones/palma/.

WRI: Development Without Conflict. (2007). Case 2. Esquel gold project, Argentina. Retrieved February 13, 2014 from http://www.sarpn.org/documents/d0002569/8-WRI_Community_dev_May2007.pdf.

Zhagüi, L. (2010). Ecuador: Cantón Girón se declara "libre de minería". Retrieved October 21, 2013 from http://www.biodiversidadla.org/Principal/Secciones/Noticias/Ecuador_Canton_Giron_se_declara_libre_de_mineria.

Chapter 5
Analysis and Interpretation

This chapter will show that the key concepts discussed in Chap. 3: relationships and their indicators, meeting transactional needs, meanings given, reference communities, and decision and change processes provide a useful framework for understanding the course of social events during mineral exploration. I begin with general comments on the relationships observed in the case studies, followed by a detailed description of the method that I used to assign a qualitative risk-of-conflict measure to each of the relationship indicators for the case studies (using the Fruta del Norte case as an example). I then combine the resulting risk-of-conflict measures for the five cases into a comparison table that presents a pictorial representation of the qualitative measures of risks associated with each case. I accompany the table with a summary risk-of-conflict narrative for each case. The remainder of the chapter discusses the interactionist processes at work in terms of meanings, reference communities and change.

5.1 Relationships

In all case studies, most interviewees considered personal relationships as crucially important, as is evident from the examples and interviewee comments shown below.

The first example refers to the mayor of a canton who delegated powers to the president of a parish based on a personal relationship, which provided the president with independence and leeway in decision-making:

…He transferred five areas of authority to me [male local authority]…

The friendship between the initial manager of the Loma Larga project and the small commune close to the mineral deposit played a big role in the subsequent establishment of good relations between the parish of Chumblín and the company:

© The Author(s), under exclusive license to Springer Nature Switzerland AG 2020   83
J. Boon, *Relationships and the Course of Social Events During Mineral Exploration*,
SpringerBriefs in Geoethics, https://doi.org/10.1007/978-3-030-37926-1_5

...our first contact was with the engineer and we became friends – afterwards the company worked with the women's organizations and after that with the parish council [male commune member]...

The problematic personal relationship between the mayor of San Antonio de los Cobres (Salta, Argentina) and mineral exploration companies prevented them from undertaking joint projects with San Antonio de los Cobres' citizens and organizations:

...the mayor of San Antonio de los Cobres charges a levy on all companies working in the area and does not want to work with us [female company personnel]...

The innumerable personal relationships of the director of Mansfield Minera with people in all social layers of society in the province of Salta, Argentina helped embed the project in local society:

...I maintain personal relations with the legislative branch, the executive branch, the judicial branch, the Church, trade organizations beyond mining... people in Tolar Grande phone me to help resolve personal problems [male]...

The personal relationships between the community relations officer of Empresa Chungar in the district of Antioquía in Peru and local citizens led to the establishment of micro-business and played an important role in the project being able to continue its work without protests or interruptions (personal observation by the author).

The personal relationships that developed between the community relations team of Minera Exar and local citizens helped create a respectful dialogue space within which the communities and the company were able to achieve mutual agreement:

...Community members confided personal matters to us, which risked creating ethical dilemmas [female company personnel]...

These are examples of personal relationships between company personnel and community members. There were indications that personal relationships between company personnel can also affect the social course of events: companies with a culture tending towards self-managed teams based on well-functioning personal relationships have an easier task of establishing relationships with communities (Boon & Elizalde, 2013).

We asked respondents to identify all organizational entities of which they were aware that existed in or affected their community (the "subunits" mentioned earlier) and to comment on the character of the relationships that existed between them. The entities most frequently mentioned in addition to the exploration company included local governance: commune; settlement (comunidad or barrio); town or parish council and its president; mayor or equivalent; the local general assembly; district, provincial and national government organizations and authorities; police; and judges of the peace. They also included institutions such as educational establishments; social welfare networks (for example, "Vaso de Leche [Glass of Milk]") in Peru; and churches. In addition, they also covered a wide range of organizations: parents associations; federations of aboriginal peoples; media outlets; drinking water and irrigation boards;

and a variety of special purpose organizations such as cattlemen, farmers, agricultural products, medicinal plants, crafts, sports and others. In many cases, the mineral exploration company and the local authorities constituted the axes that interacted with the largest numbers of other units and subunits, although in various cases other axes such as the Catholic Church or Rondas Campesinas also were important. The number of units and subunits and the frequency and nature of their interactions varied between cases. In many communities, there was a high degree of cross-membership between entities (e.g. the female president of the agricultural products society of Chumblín was also a member of the cattlemen's association and played a key role in the Catholic Church's catechesis organization). Various respondents commented that organizations are an important cohesive force in communities:

> ...It is important to maintain the organizations – they result in a stronger community...The water boards, the organizations that work on raising animals, family vegetable gardens are important for community cohesion [male community members] ...Strengthening of our organizations has been a major benefit of the project [female community members]...

This bears out this study's general observation that communities with few organizations were less articulate about their goals and in their dealings with the mineral exploration company than communities that had a more active community life. In sociological terms, the iterative processes of meaning formation and modification took quite different courses in different communities, probably because of differences in prior history, physical environment, availability and spatial distribution of resources such as water and physical proximity between community subunits. The web of relationships and interactions between subunits played a key role in the formation and change of meanings. For example, in the Loma Larga case, the establishment of relationships between the mineral exploration company and the commune close to Chumblín and with the agricultural products organization in Chumblín led to a shift of the meaning community members gave to the mineral exploration project, while at the same time strengthening organizational life of Chumblín and changing associated meanings:

> ...Through the women's organizations we have become more appreciated as women...-Machismo has decreased...The example of the women's organizations has led the men to organize themselves, for example in a cattlemen's association that also admits women as members...Opinions about mining have changed: now the majority is in favour [female community members]...

It also led to some convergence of mutually given meanings between the mineral exploration company and the community.

5.1.1 Relationship Indicators

Table 3.1 shows the relationships indicators I discussed in Chap. 3. I used information from the interviews and other sources to analyse the indicators of the relationship between the mineral exploration project and the community and to assign a qualitative risk-of-conflict measure to each.

5.1.1.1 Indicators of the Fruta del Norte—Community Relationship

In the analysis below, the relevant interviewee comments listed for each relationship indicator together with other information drawn from the literature and media sources formed the basis of a qualitative measure of the risk-of-conflict.

Trust (To believe despite uncertainty. It involves taking risk and beliefs about expected behaviours of the other)

> …Our relationship with Kinross is one of trust…Kinross has common values that coincide with what one learns as a child. The company is integrated: there is no egoism. It is a "second family" [male local Kinross employee]…Kinross has a policy of not seeing any difference between workers, geologists, technicians: all are equal and treated as equals…The different mentality of the people that work in Kinross makes that everything with the community is better managed…The relation between the president of the parish council and the management of Kinross is very much based on respect and trust, which makes that community objectives are achieved…The community will not allow a company other than Kinross to enter…I believe that they exclude the Shuar [indigenous] community from employment [male community members] …The connection should be like it was with Aurelian, there is no communication now [female community member]…

While some interviewees expressed a specific concern, most comments indicated a high degree of trust that they appeared to base on the lived values of the company.

The qualitative risk-of-conflict measure on this indicator was "very low".

Respect (To take notice of; to regard as worthy of special consideration)

> …The moral thing is very important! Kinross managers are people with values… Getting to know how Kinross is managed helped grow support for mining and for the company… The key values of the company are that people come first, and it considers differences to be strength… The company treats its workers fairly – it would prefer losing money to losing people. For example, when an illegal miner suffered an accident, they sent a helicopter to take him out of the bush… They care very much about the people of the region and take a wide range of aspects into account… They give interaction with all parties the same importance. The key value of the company is that people come first [male community members]…

The community respected Kinross, mainly because of living its values, the same as for the previous indicator.

The qualitative risk-of-conflict measure on this indicator was "very low".

Communication (Hear and being heard)

> …The community connects with Kinross on all aspects…There is active communication by the parish council or by Kinross – and they frequently convene meetings with the communities [sub-units of the Parish of Los Encuentros]…Communication with Kinross is very important for the community…We have a high-quality connection with Galo Tibbi, a Shuar who works in Kinross…Within Kinross quality of communications has been much better since Galo was appointed. People in the company are always ready to collaborate with the settlement [where the interviewee lives] and they listen to the needs of all…The parish council has a good connection with Kinross with respect to communication…The most important connections are those between the president of the parish council and Kinross management…There is a good connection with the president of the company…The president of the Shuar Federation has a very good connection with Kinross…The communication between the governor's office and the company is good and fluid [male community members]…

Not all comments were positive:

…Kinross interrupted the good relation that we initially had with Aurelian [Kinross' prede-cessor]. The connection should be as it was with Aurelian – now there is no communication. The change of personnel reduced the communication, and now there is no longer support for the community [sub-set of the parish]. In addition, Kinross only supports other communities, not the Shuar community [female Shuar community member]…

The latter comment contrasts with the comment

…Aurelian did not do things right in the past [male community members]…

While there were some concerns about changes of personnel leading to interrup-tion of established channels, the great majority of interviewees believed there was good communication. In addition, as we will see for the "focus" indicator, most inter-viewees were aware of the interaction and communication protocols that had been set up and went along with them. The comment on the effect of personnel changes confirms the assertion of symbolic interactionism theory that direct relationships are the vehicle for the social interactions that lead to meanings. Symbolic interactionism theory may help both sides understand what is going on and provide a context for rapid learning.

The qualitative risk-of-conflict measure on this indicator was "very low".

Mutual understanding (Degree to which each side can correctly express what the other side is saying)

Gauging this particular aspect reliably would have required interviews in which both sides participated. Logistics made it impossible to set up such interviews in the time available. It was therefore difficult to judge whether there was mutual under-standing, but I did encounter no evidence to the contrary. The interviewee com-ment below suggests that the company's mindset is conducive to achieving mutual understanding.

…The company cares very much about the people in the region – it takes all areas into account [female community member]…

The qualitative risk-of-conflict measure on this indicator was "very low".

Conflict resolution (Degree to which conflict resolution mechanisms exist and are productively used)

…the president of the parish council did not want there to be mineral exploitation in the zone, reason for which he imposed on the community that it reject Kinross… The Church's parish council [as opposed to the town's parish council] analyzed the situation and thought that it could contribute. The Church provided the bridge for the dialogue: as both sides trusted the priest, he could act as a bridge. Many were opposed to mining. In the beginning, things were very difficult. I had to flee and hide. Part of the problem was that some Aurelian managers did not keep their promises. There were also changes of management that caused dissatisfaction. We had many discussions and education, and we used the Internet. Once the community parish council had been established we withdrew from this effort [member of a group of male community members who saw the project as a potential opportunity]… Around 20–30% of the people of Los Encuentros are opposed to mining. They are not organized. This is "part of the game"- they are driven by fear…Los Encuentros is a quiet community and there is not much conflict… Kinross coordinates employment in the company with the president of the parish council, which makes people feel bad … This problem about employment is very damaging and generates many conflicts. The community leaders are to blame as they allowed the injustices to take place…There has only been part time work, Kinross does not give priority to the parish of Los Encuentros to provide employment, because there are workers from other cantons and the company has to give priority to people from the zone [male community members]…

The dialogue facilitated by the Catholic Church largely resolved the early conflicts around mining in the area, although some opposition remained. That opposition had not led to conflicts as it was not organized and as the majority of the population were now in favour of mining. There was dissatisfaction about the level of employment and about how the recruitment process works and who gets the jobs, but this had not led to open conflict. There was no mention of explicit conflict resolution mechanisms and it appears that the processes through which the various components of the community interacted with the parish council and with Kinross offered sufficient opportunity for resolving incipient conflicts.

The qualitative risk-of-conflict measure on this indicator was "very low".

Goal compatibility (Degree of compatibility between the goals of the parties)

Kinross' main goal was the establishment of a profitable, responsible gold mine. The "responsible" part of this goal included: sustainability; alignment with the Plan de Desarrollo y Ordenamiento Territorial de Los Encuentros (PDYOTLE— Development and Land Use Plan of Los Encuentros); added value; local development; strengthening of social organizations, one of the reasons for coordinating all actions through the parish council (personal communication, Winer Bravo).

Los Encuentros' main goal was development as laid out in the PDYOTLE and the "responsible" part of Kinross' goal coincided with Los Encuentros' development goals with the profitable mining part providing part of the required resources. The goals were compatible and both parties had agreed to them. The only barriers to achieving them would relate to implementation challenges or contextual factors over which neither party had control.

The qualitative risk-of-conflict measure on this indicator was "very low".

Balance of power (The extent to which each party influences the other and to which each party affects outcomes)

…There are power differences but we do what we can to avoid these interfering with the relationship. For the community it is most important to establish a relationship. This way you contribute to organized local development [female company employee]…The company made a mistake once: it projected an image of too much power when it offered a dinner that was too sumptuous, at least in the eyes of the community. Even though they tried to do things well, it was 'too different – a bit exaggerated'… If Kinross were to leave the community, the latter would not continue applying Social Responsibility because Kinross is always at the helm [male community members]…

Kinross had access to considerable financial resources and much technical, managerial, legal and social experience and knowledge. Los Encuentros had people with hopes and dreams, with talents, local knowledge and some education. To draw power from these attributes, they would need a strong social structure, which unfortunately was lacking. Therefore, Kinross was much more powerful and much more able to affect the outcome. This imposed a considerable burden of responsibility on the company, which it addressed through its corporate value system and CSR policy. Only one interviewee reported a mistake, and he did not consider it to have been a serious mistake.

The qualitative "score" on this indicator signals a very low probability of conflict.

Focus (Clarity about who should be legitimately involved in the relationship and about the matters at stake in the relationship)

The roles of both sides appeared to be clear and the process largely functioned as intended:

> ...The fact that the company relies on the president of the parish council to channel the resources programmed for CSR indicates that this is one of the most important connections for the company ...The interactions between the Shuar and the company are mediated by the Shuar Federation that plays a very important role. There was much vulnerability... Kinross supports joint planning with the government...Kinross is one more actor in the joint planning of development [male company employee]... The internal problems of the Shuar Federation and the subsequent change of director posed a challenge for Kinross, which stopped interactions until the Federation solved its internal problems...Before things were disorganized. Now there is a National Development and Land Use Plan. (Kinross paid for the preparation of our plan). It is a 20-year plan, revised annually. Coordination takes place up to the national level. It is a tool...The main dialogue is with Kinross managers, needs are communicated to Winer Bravo and a way of meeting the best way possible is sought, for the benefit of the community...For the company, the parish council is the most important organization...The parish council leads in the contacts with the company...The company, the parish council, the leaders of the communities [sub-units of Los Encuentros] and the volunteers are all equal...

The qualitative risk-of-conflict measure on this indicator was "very low".

Frequency (Frequency of significant interactions between the parties, whether positive or negative)

Interviewee comments indicate that there was frequent communication between Kinross and community actors, and there were no reports of negative interactions.

> ...Connections are frequent and take place at least every month and depending on the needs that arise...The requests that are put forward are proposed to the parish council and carried forward [to Kinross] by it – Kinross is nimble and not much bureaucracy is involved...Requests are frequent, in accordance with the needs that turn up. The community has quite a few needs...The frequency of the communication with Kinross that exists depends much on the needs that show up. Lately there have not been contacts...There are meetings when requested [male community members]... With the parish council there are permanent formal and informal meetings – the relation is very fluid. There are many links...With the Shuar Federation there are a minimum of two meetings per month – the meanings are conducted in Spanish...We work together with all actors. With critics [of the company] at least once every 15 days [male company personnel]...

According to Gawley, establishing routine activities such as regular meetings serves to "sustain the acceptance of trust definitions" between actors. It also helps establish new social structures and solidify the relationship (Gawley, 2007).

The qualitative risk-of-conflict measure on this indicator was "very low".

Stability (Degree of predictability)

Some of the interviewee comments made in relation to other indicators referred to company personnel changes that affected the relationship. Most related to the changeover from Aurelian to Kinross, and Shuar made most. None of the other interviewees commented on this matter, which was probably specific to a particular group and situation. There were no reports of recent serious conflicts or significant changes in the relationship, and it appears that the interactions between Kinross and the community had settled into a pattern familiar to both and that excursions from the pattern were manageable.

The qualitative risk-of-conflict measure on this indicator was "very low".

Productivity (Degree of achievement of target results)

> … The relationship of the company with the community with respect to education is very good, as it created the condensed basic education program that was an incentive to study more… The different mentality of the people working in Kinross contributed to better management with the community [male community members]…First know the community, what are its needs, why they have problems. Money has no value – accompaniment is what is needed. The operational plans of the Shuar Federation were developed with the accompaniment of Kinross…The young people are ashamed to speak Shuar and to dress like Shuar. They [Kinross] have organized a meeting between Shuar communities to promote the use of the Shuar language – the young people will have value from this [male Shuar community member]…The connections with Kinross are very good because they look after our needs, not all of them, but they try to meet the majority even though the community sometimes has to push…The arrival of Kinross was very beneficial to the community because it created a source of employment…In addition they much supported education through which community members obtained their high school diplomas, and they supported the creation of small enterprises…I am in agreement with the arrival of the company, because it benefited me in the creation of my new employment…The company has carried out infrastructure in the community, which has also been a very good contribution to the community…The company does indeed care for the people in the communities and it has carried out public works that benefit them. Even though the communities have many needs, the company has always tried to help us [male community members]…Kinross has not supported us much – the Shuar community was better off with Aurelian…The majority of the proposals made by the company have not been carried out [female Shuar community member] …

While most interviewees were of the opinion that the relationship had been productive for the community, there were some dissenting Shuar voices. The section on the perceived benefits and harms in the case study (Chap. 4) indicates good progress towards the development targets of Los Encuentros and the relationship targets of Kinross. Unfortunately, Kinross and the national government did not reach agreement on royalty issues and Kinross decided to withdraw from Ecuador. As was mentioned earlier, Lundin, a Canadian company that was well received by the people of Los Encuentros, purchased the project.

The qualitative risk-of-conflict measure on this indicator was "very low".

5.1.2 Relationship Indicators Compared Between Cases

I derived risk-of-conflict measures for the indicators for the other cases in the manner I just described. Table 5.1 summarizes the results for the case studies described in this book. The fill colour of each cell corresponds to the measure of conflict risk for that particular case study and indicator. For the cases surrounded by a mix of communities some of which were in favour of the project and some against, I took into account the perceived intensity of opposition and its potential impact on the project. The table allows differentiation of the cases at a glance. For each case, it also gives a quick indication of the aspects of the company-community relationship that contributes most to the risk-of-conflict and that would need the most attention to avoid conflict.

The discussion below explains the reasons behind the assignment of conflict risk measures. It draws on the summaries of the individual case studies in Chap. 4.

In the Palma case, the factors that negatively affected the relationship indicators include the generally low level of trust between citizens in the area (Vargas Gonzales, 2013) and the bribery scandal surrounding Minera Huascarán (the predecessor of Empresa Chungar). In addition, the relative newness of Volcan Minera to social responsibility in its operations (company culture sometimes got in the way, and the community relations representative had to work with limited resources).

Table 5.1 Summary of qualitative risk-of-conflict measures for company-community relationship indicators

Very low	Low	Neutral	Moderate	Medium	High	Very high

Indicator / Project	Trust	Respect	Communication	Mutual Understanding	Conflict Resolution	Alignment	Balance of power	Focus	Frequency	Stability	Productivity
Palma											
Loma Larga											
Fruta del Norte											
Cauchari – Olaroz											
Lindero											

The level of risk is colour-coded as shown in the bar below. Blank cells indicate that insufficient information was available

In the Loma Larga case, there was strong and sometimes violent opposition by the parish of Victoria del Portete, less strident opposition by the authorities of the canton of Girón (in which the mayor's political identity and associated reference community played a significant role in establishing meanings and choosing options for decisions). The mayor of the canton of San Fernando (who gave a different meaning to the project than what he perceived to be the negative meaning assigned to it by the majority of the citizens of the town of San Fernando, the capital of the canton) was "neutral".

Both Victoria del Portete and the town of San Fernando are outside the zone of the direct influence of the Loma Larga project. Note that the parish of Chumblín that belongs to the canton of San Fernando and the parish of San Gerardo that belongs to the canton of Girón came to see the arrival of the project as an opportunity through a process that I will describe later in this chapter.

In the Fruta del Norte case, the balance of power tilted heavily towards the company. However, the company was fully aware of the power imbalance and was bending over backwards to prevent it from negatively affecting the relationship.

The Fruta del Norte, Cauchari-Olaroz and Lindero cases had the lowest overall conflict risk assessments and had not seen serious conflict at all.

In summary, personal relationships between actors and relationships between groups of actors exercised an important influence on the course of social events. Assigning a qualitative risk-of-conflict measure to the relationship indicators clearly differentiated the cases and identified areas that need attention from the communities and companies.

The section that follows discusses the part relationships play in the interactionist processes and the role of these processes in bringing about change, for each of the case studies.

5.2 Meanings, Reference Communities, Relationship Patterns and Change Processes

5.2.1 Palma

The question on reference communities that I added to the interview guide for this case study was "When you make a decision, from where come the values on which you base the decision?" The citations below indicate that there is a range of reference communities on which people draw. However, for the majority of interviewees the community to which they belong also served as their reference community, and the general assembly played an especially important role in this respect.

> ...I try to capture the orientations of the people that surround me every day [symbolic interactionism uses exactly the same concept]...The community...The comuneros...The citizens...The community through the assembly...The assembly...Assemblies...The general assembly...The president guides the assembly...Being a comunero ⇒ the community

decides ⇒ the comuneros have a common interest ⇒ this would be the reference community…We need the others to decide [exactly the same as the symbolic interactionist
concept]…The community, with the help of professionals in it…Professionals in the community…I am a Christian…I learn from Christianity and my education… The Bible…The
Incas: one for all and all for one…The community of judges is my reference community…A
person who knows more than I do. Most comes from outside… Older, respected people…-
Much has changed over the past 20 years. In the past people were more respectful. Because
of modernity the young are no longer respectful…It is not difficult: values of upbringing,
grandparents, parents [male community members]…

The general assembly was the most important mechanism "of interaction that one
has with one's fellows from which the meaning of people of things is derived", and
"the interpretive process through which meanings are managed and modified when
a person deals with the people or things that he or she encounters". As postulated by
the theory of symbolic interactionism, this involves direct relationships. The general
assemblies are extremely important in terms of formation and modification of meanings. At the same time, they constitute a reference community. A number of interviewees also called on various types of authorities or on professionals inside or outside
the community for help when they encountered challenges. Some interviewees mentioned religion and contrary to what might be expected, relatively few mentioned
the values of their upbringing, their grandparents and their family. One expressed
concern about what he perceived to be the decline in traditional values among youth.
This may express a "normal" generational gap that occurs in many societies or there
may be more to it. Unfortunately, it was not possible to verify this assumption through
additional interviews with youth. Youth and adults appear to draw on quite different
reference communities. Claverías Huerse and Alfaro Moreno assumed that education is responsible for the difference (Claverías Huerse & Alfaro Moreno, 2010).
This suggests that education affects youths' reference community, a subject worthy
of future research. Not all interviewees understood the interview question and future
studies will need to refine the approach. The above description shows a complex,
tightly woven web of relationships and personal interactions. These were played
out through the governance structure of the comunidad campesina and mechanisms
intertwined with cultural and economic practices, tradition, and were conditioned
by geography and climate (the river provides prawns and irrigation water, the soil
is fertile). In terms of symbolic interactionism, these frequent and intense personal
interactions would lead one to expect that meanings develop rapidly. However, as
the general level of trust in the area was low the interactions, although frequent, may
not be positive and this slowed down the evolution of common meanings. Claverías
Huerse and Alfaro Moreno's observations seem to confirm this assumption.

Incidents similar to the Huascarán corruption scandal would have led to demon-
strations and protests and could have resulted in violence in other areas (for example
in Piura, see Boon, 2015). Why was there a different reaction in the Lurín Valley?
A plausible explanation is that the general low level of trust was a barrier to the
formation of an interpersonal network that could lead to a convergence of meanings
around which a large-scale group identity could be constructed (Deaux & Martin,

2003). As there was no convergence of meanings, there was no verification of a common interpretation of the situation and a group decision on action was not possible. In the context of the many reservations, they had about the Palma project and its parent company, why did many interviewees comment that they have good, trusting relations with the Palma project's community relations officer? In interactionist terms, through his systematic, frequent interactions with community members he had over a period of three years developed his meaning of the communities, and community members had developed their meanings of him and the latter converged to some extent. However, they gave this meaning to him as a person and it only weakly reflected on the company. Even those in Espíritu Santo who were opposed to the company respected him, albeit somewhat grudgingly.

5.2.2   Loma Larga

In the Loma Larga case, members of the anti-mining camp focused around Victoria del Portete met regularly:

> …We see each other every three or four days to exchange information and build mutual trust…Our relation with CONAIE (Confederación de Naciones Indígenas del Ecuador – Confederation of Indigenous Nations of Ecuador) is the most important because we are Cañari [male members of the group opposed to mining]…

These regular encounters were part of the process of establishing meaning, interpreting the situation and contemplating action. With respect to the reference community on which this group drew the series of photographs taken of speakers during the public consultation held in Victoria del Portete shown on Flickr depicts an important part of their reference community—a concept made visible! (Flickr, 2011). A similar process occurred within the company to arrive at a decision on how to deal with the communities. The reference communities and the identities of each of the parties formed part of this process. The processes in the community likely were much more diffuse than the processes in the company. According to Ashforth and Fried, within organizations "…the bounded and ongoing nature of action, and the interdependence of organizational actors both necessitates and facilitates the routinization of expectations and behaviours, that is, the development of organizational scripts". Their work indicates that processes within the company likely have structure and operate at greater speed (Ashforth & Fried, 1988).

The relationship patterns in the parishes of Chumblín and San Gerardo included a number of organizations and associations. In both parishes, there were both people for whom the arrival of Iamgold meant an opportunity and people for whom it meant a threat. The interactions in San Gerardo took place in a context of trusting and respectful relations between the company, the parish council, its president, organizations and associations and community members. This resulted in a decision to work together with the company:

...We have excellent communication and consult with all levels of the company...The company is prepared to cooperate in development and is considered a member of the community...The population, the local government, the company work together in a transparent way...We do everything together [community members]...

In the area of Chumblín, the commune saw the arrival of the company as an opportunity, the establishment of personal friendship between a manager of the company and the commune being an important ingredient of the process as I mentioned earlier. The interviews in Chumblín showed that the company and the groups that saw the arrival of the company as an opportunity established trusting and respectful relations:

...With the company we organize into productive groups that sell their products in the markets... 'La Natividad de Chumblín' produces agricultural products...We have good relations with both Chumblín and the company [a male and a female member of the commune, interviewed together]...We have a strong and productive relationship with the company. As an association, we are independent and not beholden to the company...Iamgold (predecessor to INV Metals) has made an important contribution and was the first company that paid attention to us [female community members]...

At the beginning of the process, those who saw the arrival of the project as an opportunity formed a minority, and their decision to undertake joint projects with the company landed them in hot water on many occasions:

...Our husbands did not let us into the house after we had been to a meeting of the association...Some husbands beat us up and left us with blue eyes...They accused us of 'selling out the parish'...[Other parish members] insulted us...We trusted the company and persevered...It was very difficult and there were ugly moments. This went on for some two years [female community members]...

However, they persisted and when they shared the results of their labours, they changed the meanings others gave to the company. Through relationships and associated interactions, the meanings given by individual members who initially had doubts about the arrival of the company converged towards acceptance and opportunity.

...In the past there were certain disagreements between organizations. We did not know what mining was. The company's socialization led to discussions within and between organizations. It took four to five years and now 85% of the population agrees with mining. The company has been a unifying force...Many did not realize that the company could help [male community members]...

Both the company and the communities invested considerable effort in establishing these relations and the majority of respondents were of the opinion that this effort paid off:

...People know each other. They have gained trust and support and benefited economically. From an organizational aspect we learned how to coordinate, how to maximize the benefit from the support, which is a benefit for the future...Our personal life has improved...Chumblín is much more united now [male community members]...

The company facilitated the establishment and maintenance of relationships through a capable and committed community relations team that was working jointly with the communities. In turn, the communities working together strengthened their organizations and signed agreements with the company. This showed overlap that developed between the meanings and reference communities of the communities and of the company, that is to say, a common space had formed. The community became part of the company's reference community to a certain extent and vice versa.

Through the dynamic and recursive processes that I mentioned earlier, meanings, reference communities and relationships evolved. This led to the signing of agreements between the company and the parish council in both San Gerardo and Chumblín and to the results referred to by the interviewees. The reference community of the presidents and of the people broadened and now probably includes aspects related to the company while the reference community of the company now includes elements related to the parishes.

The local relationship patterns and their characteristics in the parishes of San Fernando, Girón and Victoria del Portete were probably similar to those of Chumblín and San Gerardo, but their reference communities were different. As I mentioned earlier, these communities were anti-mining, their reference communities may be environmental organizations. The reference community of Victoria del Portete may be the leaders that spoke during the public consultation exercise, that Flickr showed (Flickr, 2011). Nevertheless, the processes described above led these parishes to give the meaning of a threat to the arrival of the company and relations and meanings evolved from then on to the present point at which these actor groups remain opposed to mining although the actions they decided to take differ between them. Either these groups had no relationships with the groups that favour mining, or their relationship was one of strong animosity. This means that it was difficult for either side to influence the meaning given to mining by the other. There were differences between the groups opposed to mining and interviewees suggested that members of these groups might be ambivalent about the meaning of the presence of the company:

> …A survey conducted by Ecuador Estratégico in the area of Tarqui and Victoria del Portete suggested that a high percentage of respondents were in favour of mining…Over the past year I have detected a shift towards acceptance in Victoria del Portete [local female company employee]…

As I noted earlier, I was not able to obtain a copy of the survey mentioned by the interviewee or otherwise to confirm its results. In the case of the groups of actors opposed to mining such as the parish of Victoria del Portete, the difference between their reference community and that of the opposite side increased with time.

5.2.3 *Fruta del Norte*

According to respondents and to the Plan de Desarrollo y Ordenamiento Territorial de Los Encuentros (PDYOTLE—Los Encuentros Development and Land Use Plan—Los Encuentros is the parish closest to the project site), the arrival of Aurelian

Ecuador caused great inconvenience from a political, social, economic and environmental perspective (Gobierno Parroquial de Los Encuentros, 2011). An interviewee comment confirmed this:

> ...It should be mentioned that president of the parish council did not want there to be mineral exploitation in Los Encuentros, for which reason he imposed on the community that it reject Kinross [male community member]...

In interactionist terms, the meaning that people gave to Aurelian Ecuador through their interactions with their fellows led them to interpret the arrival of Aurelian Ecuador as a threat. People then drew on their reference community to decide what action would fit with its norms, and community leadership decided to reject Aurelian Ecuador, which probably also included instructions to avoid direct contact. Because detailed probing of the past was beyond the reconnaissance scope of the present study, I could only make assumptions about how community members arrived at the particular meaning they gave to Aurelian Ecuador and what their reference community's norms looked like. Had the example of Nambija influenced its norms related to mining? (Nambija is a close-by lawless town where artisanal and small-scale illegal mining have caused huge environmental damage as well as landslides that killed hundreds of people and that suffers from serious social problems). Alternatively, did its norms relate to preservation of the agricultural lifestyle? Or, was it that, rather than fear being inspired by mining, there may have been a fear that the arrival of the company would result in the banning of artisanal and small-scale mining (illegal or not), an important source of income in Los Encuentros (Agreda Orellana, 2011).

As was mentioned earlier a small group of people linked to the Catholic Church in Los Encuentros gave a different meaning to Aurelian Ecuador: they did not see it as a threat and therefore decided to talk with the company, i.e. they "dealt with the thing they encountered" and in the process, they "handled and modified its meaning". They began building a bridge between Aurelian Ecuador and those that saw it as a threat. In the process, the latter began handling the initial meaning they gave to Aurelian Ecuador and modifying it. The process was slow and full of twists (at one time the "bridge builders" had to flee and hide for a while). However, the process led to the parish council's decision to work with the company. While this process was going on in the community, presumably the "bridge builders" also allowed Aurelian Ecuador to "handle the meaning it gave to Los Encuentros", modify it and change its interpretation of the situation. At the same time as Aurelian Ecuador was changing its interpretation of the situation, the norms of its reference community were changing also. Over the past decade, interest in Corporate Social Responsibility skyrocketed and community engagement started becoming the norm (Boon, 2012). As the norms of its reference community changed, so did Aurelian Ecuador's decisions and approaches, which possibly facilitated their local policies. The latter, in turn, helped change the meaning given to the company by the community. According to respondents quoted earlier under the Fruta del Norte "conflict resolution" indicator, the fusion of Aurelian Ecuador with Kinross caused a further modification of the meaning each side gave to the other. From the perspective of interactionist theory, the culture of Kinross represented a reference community with clear norms

(Kinross values) against which employees measured decisions and actions and, very importantly, the community was aware of and appreciated these norms. In effect, Kinross culture became part of the community's reference community. At the same time, the close cooperation between the parish council and Kinross was leading Kinross to incorporate community norms into its decision-making processes, thereby making the community part of its reference community. Kinross' value system and the comportment of its employees clearly had a strong influence on the meaning interviewees attached to Kinross, as is apparent from their comments that I cited earlier under various relationship indicators. It appears that the company's internal "…processes posited by symbolic interactionism led to a collectivity and a common meaning: norms and expectations evolved concomitantly to regulate and maintain the collectivity" (Ashforth & Fried, 1988: 320). Scripts were successfully used and Kinross avoided the "pressures toward automatic behaviour" that can engender "…the indifference of many organization boundary spanners in roles where 'authenticity' is expected [as in interactions with communities] that reflects an affective as well as a cognitive detachment" (Ashforth & Fried, 1988: 316).

As said earlier, meaning results from social interactions and the relationships, interactions and interpretations at work in CSR situations constitute the dynamo that drives the unfolding of events. The relationships of the "bridge builders" with the community and with Aurelian Ecuador allowed them to start a change process. Respondents' comments indicate that the direct relationships with Kinross executives and employees very much influenced the meaning respondents attached to Kinross. The importance of personal relationships in determining the course of events was mentioned a number of times as well in the context of bilateral relationships other than those involving Kinross, and I quoted a number of such comments at the beginning of this chapter.

5.2.4 Cauchari-Olaroz

The communities gave an initial nuanced meaning to the Cauchari-Olaroz project:

> …Some communities were opposed and some thought that it would be worth having a discussion. Through a discussion process that took between one and a half and two years, the six communities of the Department of Susques jointly decided to work with the company [female community member]…

Through their interactions, the participants converged around a common meaning and interpretation of the situation and arrived at a decision. The reference communities on which they drew to arrive at this decision were mainly the general assemblies:

> …Minera Exar has always worked together with us and for this reason people have no problem with them, especially because matters are talked about in the assembly. …They inform us about their plans and we want them to proceed. …They are responding to our worries and well…The company always consults us through the assemblies [male community members]…

Real and imaginary members of the reference communities of community members likely included general assemblies, other communities in the zone, tribal organizations, umbrella indigenous organizations, spiritual leaders, oral traditions and possibly certain NGOs. This aspect needs more study. A female member of one of the six communities made a presentation during a social responsibility seminar in Salta, Argentina. She melded the ideas and concerns alive in her communities with norms on indigenous issues being articulated and promoted through organizations such as the International Labour Organization, the United Nations, the International Finance Corporation and others to explain why the communities had decided to support the company and what safeguards they were putting in place (Calpanchay, 2012):

> …These change processes are above all linked to the right of the communities to compulsory prior consultation on the implementation of mining activities on their lands, to determine if their interests will be affected and/or harmed by the development of a mining project that carries possible environmental, social economic and cultural impacts… Now we bet on this activity, generating change and if we make mistakes, we will look at the point where we made the mistake to correct it. We do not want nor will we permit that others determine our future. Rather we invite them to take part in the sustainable development that comes hand in hand with respect…Now the company is a new neighbour joining us and we open the door to them, but without forgetting that we will be the first supervisors of the precepts on which we agreed and if we see some mistakes we will resort to the dialogue that characterizes us to ask for the corresponding explanations and rectifications…

The latter comment about rectification could have come straight from the United Nations Guiding Principles on Business and Human Rights (Ruggie, 2011). However, she did not make any direct reference to these international norms: they had become part of the broader global indigenous reference community. Aspects of this reference community overlap with those of the professional and peer reference communities of the company (both probably drawing on some of the same sources), which definitely facilitated reaching an agreement.

With respect to the reference communities drawn on by company staff and management, the most relevant way of categorizing subsets within the company was by "employee origin" and by "rank", leading to the subsets: "local employees"; "non-local employees"; and "management".

Current members of the reference communities of management likely included company head office, relevant professional organizations and peers in mineral exploration and mining. At the time of the study, the "company" reference communities were probably nearly identical to the "management" reference communities. Company interviewee comments that confirm some of these assumptions include:

> …Our knowledge of e3 Plus came after we initiated our RSE practices based on the Equator Principles. Based on that, we developed a code of conduct that embraces policies for social and environmental practices… E3 Plus helps us fill the gaps in our social programs and improve our environment and occupational programs. It can help us develop a more defined, stronger management policy to create managerial capacity building so that the managers can transmit these skills to their employees and try to prepare for the transition from exploration to development. This is a big challenge. It would be good if e3 Plus were translated into Spanish… Minera Exar adheres to the Equator principles (sustainable economic growth;

environmental protection; well-being of the people who live close to the project or that can be affected by our operations)…For the CEO CSR is as important as the rest of the business. Personnel follow CSR policies well…The company recognizes and respects indigenous peoples' rights as defined in ILO 169…. The communities' rights to be consulted and access information are respected and a permanent two-way communication is maintained. Community approval is sought before starting the various phases of the project… In addition to e3 Plus, Minera Exar is guided by ISO 26000, 14000 and 9000. The government wants to implement ISO 26000… The comparison of the projects in different countries will help us create synergies to guide us in certain situations and solve them – it is not easy… In Health and Occupational Safety we have already made some progress but in the social area, as it is new, we are still exploring how to do things well [comments by male and female employees made during a group meeting]…

This will evolve with time, as by nature, relationships are not static. In adopting the Equator Principles and later e3 Plus for its social responsibility approach, company management was measuring itself against the norms of professional organizations and peers, and the incentive for doing so came from the CEO's experience with a "social responsibility disaster" in a previous project:

… I noticed the problems the Esquel project in Chubut had with communities, and how these problems forced the company to close down its operations [male CEO]…

The "peer community" aspect was evident from a certain rivalry in social responsibility matters between Sales de Jujuy and Minera Exar, albeit in terms of some "one-upmanship" rather than convergence towards common norms.

Local employees' reference communities likely consisted of the local community members and leaders; the general assembly; religious and spiritual leaders; and family networks, some of which were evident from a group of employees' decision to approach the community leader for help with solving a problem at work:

…We had a problem with a group of employees who complained about how they were treated by a certain manager to the leader of their community rather than to management. The community leader talked to management and the problem was addressed [male manager]…

and from employee concerns about the company not offering to the PachaMama (the earth/time mother goddess):

…Our hopes for the future include… that our culture be respected and that they participate in the Pacha Mama… Certain days are sacred to us (first of August – Pacha Mama Celebration) but they do not give us permission. There is a belief that working on August 1 can lead to the earth punishing you. People of our level were not working but we felt bad about it. We need to talk to the managers so that they understand this is important for us – it is a custom that comes from our ancestors. In my community, they ask if a "convido" (an invitation to mother earth) is made especially in the salt flat where the earth is rich in minerals. They ask me how could they forget about the earth? [male local employees]…

At the same time, several local employees were well aware of and influenced by company norms and values, as part of the dynamic change process:

…Now we are like a child growing up with the values of the company. Seeing the good example of our superiors, we know what to do when we grow and continue as a larger company… CSR has a significant impact on my day-to-day work. Although we do not have a course yet, I want to learn about it and apply it. CSR is becoming more and more important to the managers [male company personnel]…

The subset of non-local employees (mostly from other provinces of Argentina) probably did not have a single group of reference communities: each related to his or her reference community "of origin". With time and continuing interaction, "the company" may become more and more included in the company subsets' reference communities.

In summary, the initial meaning some communities gave to the company was that of a threat while others sensed an opportunity. Through a process of intense interaction, the communities converged around seeing the arrival of the company as an opportunity. In doing so, they drew on the norms of both local and extended reference communities. The company drew on some of the same extended reference communities and through ongoing company-community interactions, the community and company meaning sets began to overlap. The community members who were company employees mediated part of this process. This process of change is ongoing.

Because of financing difficulties, Minera Exar had to discontinue operations for more than a year. However, activities resumed late in 2014 and the relationship between the company and the communities continued developing along the path that was set out initially (personal communication, Rosana Calpanchay and Mónica Echenique).

5.2.5 *Lindero*

The interviewee comments shown below paint a picture of the company as an organization in which interactions take place through a well-connected network of fluid high-quality interactions.

…Vancouver head office fully trusts our local office and gives it great freedom of action; at the same time, the boundaries are clear and sufficiently broad; the company is thorough and patient… Management decides on most CSR issues and actions. Everyone is aware of these decisions and they are taken into account in logistics, purchasing and other company activities. There is open communication about all this [male company management]…

…We are geology assistants and undertake prospecting on our own, spending 7–12 days consecutively in the field, mostly on provincial lands. We have learned how to prospect while working for the company… We are the eyes and ears of the company and when we see or hear something that falls outside our (wide) range of responsibility we advise the director who then takes action at a higher level…We like working for Mansfield because we know what is expected and we are left free to decide how we want to go about our work…It is part of our duties to contribute to the community… I worked in a bigger company before that was more rules-bound and bureaucratic and I like my present job much better [male company personnel]…

…The work these people do is trusted completely and the reason for which they are sent out to prospect on their own – they played an important role in the discovery of the Lindero and Arizaro prospects…New employees usually ask for a detailed job description. They are told that, while everyone has a major area of activity, they are also expected to pitch in where necessary e.g. take part in the vaccination of llamas, the cleaning of the buildings – regardless of their function or level of education [male company management]…

...Daily work consists of prospecting, camp maintenance, work on airstrip construction, and any other duties that may be needed. We work as a team and rotate tasks based on mutual understanding. The company is very flexible both in terms of letting us decide how we want to approach our tasks and giving us time off when the need arises (e.g. illness in the family, feeling ill at ease in the camp, etcetera). We handle the effect of such absences on our workload as a team. We all belong to nearby communities and often meet with community members during our work. We routinely give community members rides and when other requests are made or problems crop up we usually talk them over as a group. The group leader decides how to approach the situation based on the discussion. There have not been any significant problems that we can remember [Group interview in the field camp] [male company personnel]...

These observations paint a picture of the company as an organization in which interactions take place through a well-connected network. The characteristics of the internal relationships as discussed earlier suggest that interactions between actors will transmit ideas, emotions and content fast and well. Symbolic interactionism would predict that this, coupled with the connectedness of the network, augurs for rapid establishment and convergence on meanings that are consistent between actors. The company clearly serves as a reference community for its employees, who measure their decisions against its norms: teamwork, the importance of relationships (linked to a subset of norms described by their indicators) and the concept that everyone is important. Employees adhere to these norms implicitly and semi-subconsciously.

Because almost all employees were local, or at least from the province of Salta, they shared meanings that were related to the reference communities of their non-company identities, i.e. the reference communities of the company and of the local actors "met" in the company's employees. This, together with the relationship and interaction skills that are evident from the indicators, facilitated the convergence of meanings between communities and the company. In addition, the company took a long time to bring the project to fruition, which also facilitated the convergence of meanings and reference communities:

...We have been working in Salta since 1994, and the Lindero prospect was discovered in 1999 [the author visited Mansfield Minera's office in 1998][male company manager]...The municipality has regular civil defense meetings in which all exploration companies active in the area participate. These meetings have resulted in companies putting in additional roads, improving road signage and emergency planning (for example when German tourists became stuck in the sand far away from Tolar Grande...We are aware that the transition to a mine will bring many changes and the municipality is planning for this [local authority]...During our weekly runs to the city of Salta we always carry community members and as a result we have regular eight-hour conversations with a large proportion of the community [male company personnel]... I had a chance to work with another company but I waited till there was an opening in Mansfield Minera [female local employee]...The company CSR manager will take the man who lives in the oasis close to the camp to Salta to help him take care of his pension arrangements – he is illiterate ...The company has played a key role in ensuring that the exploration companies active in the area coordinate their CSR approaches [company manager]... All actors are important. We know everyone and everyone is important... We feel part of the community and talking usually resolves any problems. On a number of occasions the community itself resolved problems, for example when some people said certain things about the company and others didn't agree, community members resolved the matter amongst themselves and it was settled [male local company employees]...Sometimes community members call me about community issues even before they call the intendente of Tolar Grande [male company manager]... The relationship between the companies and the municipality is good and there is an open dialogue... There is not much conflict in Tolar Grande and most issues get settled in a "natural" way [male company personnel]...

The above series of interviewee comments suggest that Mansfield Minera functioned as a dynamic network of actors that seemed to interact almost seamlessly with other networks of actors. Interactions through patterns of relationships drove the functioning of each of these networks (i.e. assigning meanings, interpreting situations and making decisions), and the networks "scaffolded" onto the norms of the related reference communities. The reference communities of the company were the community of Tolar Grande (including its cacique and intendente), the Salta Chamber of Mines, head office in Vancouver and its own employees. It also fitted into the important catholic religious base of the Province of Salta and supported indigenous beliefs:

...I am godfather to the daughter of one of my colleagues and I take my duties as a godfather very seriously...We take part in the annual Virgen del Milagro pilgrimage to Salta and in indigenous festivals [male company personnel]...

The company had become fully part of the community.

The question remains as to why this happened in Tolar Grande and not in San Antonio de los Cobres. The interactions that lead to establishment and adjustment of meanings and of reference groups have a random aspect and outcomes are contingent on boundary conditions. The refusal of the mayor of San Antonio de los Cobres to allow companies to undertake social initiatives was a boundary condition that reduced the number of solutions to zero. The change of mayor that took place after completion of the case study changed that boundary condition with immediate consequences: as mentioned earlier the new mayor hired Mansfield Minera's CSR manager to assist him with social responsibility initiatives presumably in cooperation with exploration companies, and productive cooperation resulted.

Perhaps not surprisingly, the most important type of reference community for members of the local populations in the Andes was local institutions: the town council or its equivalent and the general assembly. For companies, it was their peers in industry, through industry associations and the general background of CSR codes and guidelines. The processes of changing meanings all took place through interactions mediated by relationships and in a number of cases a change in or expansion of reference communities accompanied the process. In "ideal" cases (from a conflict avoidance perspective), both company and community actors began including "the other side" in their reference community, and in the Lindero and Cauchari-Olaroz cases, the company came to be considered as part of the community. In cases in which the two sides had completely different reference communities and in which the nature of the relationships between the two sides precluded interaction (i.e. transactional needs were not met at all), meanings either did not change or became more polarized (Río Blanco, Victoria del Portete).

References

Agreda Orellana, W. G. (2011). *Inventario de las necesidades básicas insatisfechas y conflictividad social de las comunidades del área de influencia del Proyecto Estratégico Nacional Zarza, Provincia de Zamora Chinchipe, Cantón Yanztaza, Parroquia Los Encuentros.* (Magíster en Gestión y Desarrollo Social, Universidad Técnica Particular de Loja), pp. 1–62.

Ashforth, B. E., & Fried, Y. (1988). The mindlessness of organizational behaviors. *Human Relations, 41*(4), 305–329.

Boon, J. (2012). Un nuevo enfoque de la responsabilidad social empresarial. *Minería Online, 418* (Especial Agosto 2012 Medio Ambiente), pp. 54–60.

Boon, J. (2015). *Corporate social responsibility, relationships and the course of events in mineral exploration—An exploratory study* (Unpublished Ph.D. thesis). Carleton University, Ottawa. https://curve.carleton.ca/system/files/etd/6c6598d4-c436-409e-9ba1-40dea2d37d2c/etd_pdf/7b39ca613ff7e7e2df52ed82580e3974/boon-corporatesocialresponsibilityrelationships.pdf.

Boon, J., & Elizalde, B. (2013, Septiembre). Relaciones: Un elemento esencial de la responsabilidad corporativa en exploración minera. *Minería, 432,* 161–173.

Calpanchay, R. (2012). Comunidad aborigen de Puesto Sey "Termas del Tuzgle". Retrieved February 10, 2014 from www.olami.org.ar.

Claverías Huerse, R., & Alfaro Moreno, J. (2010). *Mapa de actores y desarrollo territorial en la Cuenca Lurín.* Lima, Perú: Centro de Investigación, Educación y Desarrollo.

Deaux, K., & Martin, D. (2003). Interpersonal networks and social categories: Specifying levels of context in identity processes. *Social Psychology Quarterly, 66*(2), 101–117.

Flickr. (2011). Sistema comunitario agua potable VP-T. Retrieved October 21, 2013 from http://www.flickr.com/photos/62081634@N05/6215588294/in/photostream/.

Gawley, T. (2007). Revisiting trust in symbolic interaction: Presentations of trust development in university administration. *Qualitative Sociology Review, 3*(2), 46–63.

Gobierno Parroquial de Los Encuentros. (2011). *Plan de Desarrollo Y Ordenamiento Territorial - "Resumen Ejecutivo".* 2011: Gobierno Parroquial de Los Encuentros.

Ruggie, J. (2011). *Report of the Special Representative of the Secretary-General on the issue of human rights and transnational corporations and other business enterprises, John Ruggie—Guiding principles on business and human rights: Implementing the United Nations "Protect, respect and remedy" framework.* Retrieved June 28, 2012 from http://www.business-humanrights.org/media/documents/ruggie/ruggie-guiding-principles-21-mar-2011.pdf.

Vargas Gonzales, S. (2013). *Plan de desarrollo agropecuario de la parte media y alta de la Cuenca de Lurín 2013–2018.* Lima: Centro Global para el Desarrollo y la Democracia/Mancomunidad Municipal Cuenca Valle de Lurín.

Chapter 6
Generalized Model

The analysis and interpretation of the interviewee comments and literature and media sources presented in the previous chapter showed that, even though the cases are sui generis, there are many similarities between the types of processes at work. The present chapter combines these observations into a generalized model that shows how these processes lead to change, the establishment of new relationship patterns and social structures, and the course of social events and perceived present and future benefits and harms. It also shows where additional factors such as company and community characteristics, time and contextual factors such government policies exert their influence.

The model consists of a series of stages with additional influences feeding into it. Figure 6.1 shows a schematic representation of the model. There is no clearly marked boundary between the stages: they flow into one another and depending on circumstances, any stage can influence any other stage. The arrival of the project (stage 1), the establishment of initial meanings, interpretations and decisions (stage 2) and the meeting of transactional needs (stage 3) lay the ground for building relationships (stage 4) that is at the centre of the model. The relationships that are built influence subsequent change processes (stage 5) that lead to the establishment of relationship patterns and new social structures (stage 6), which in turn lead to the course of social events and perceptions about benefits and harms (stage 7). The iterative loop: (3) meeting transactional needs $\Rightarrow$ (4) building relationships $\Rightarrow$ (5) change processes $\Rightarrow$ (3) meeting transactional needs is the model's "pulsating heart" that ensures continuous adaptation of the actors to each other and to the additional factors community characteristics, company characteristics, contextual influences and time. In the context of the model, "building relationships" has no normative content: depending on what happens during stages 1–3 the relationships built can range anywhere from negative to positive. The following sections provide a more detailed description of the stages.

© The Author(s), under exclusive license to Springer Nature Switzerland AG 2020
J. Boon, *Relationships and the Course of Social Events During Mineral Exploration*,
SpringerBriefs in Geoethics, https://doi.org/10.1007/978-3-030-37926-1_6

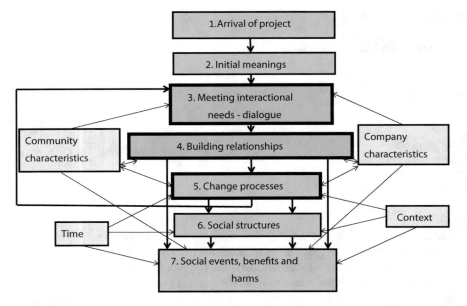

Fig. 6.1 Generalized model of sociological processes during a mineral exploration project

6.1 Stage 1: Arrival of the Exploration Project

"Arriving" can occur in a variety of ways: a company can show up one day with people and equipment and learn "on the go" or it can take a carefully planned approach
involving the previous study of relevant documents, identification of actors and gradually building up a presence. The latter would already include some elements of
subsequent stages. In all Latin American case studies, the arrival of the exploration
project was a significant event in the life of local actors. Reasons for this include
the considerable difference in culture between the new arrival and the local scene, a
generalized fear of mining that exists in many areas of the continent, and the signifi-
cance of the people and financial resources of the company as compared to those of
local actors.

6.2 Stage 2: Initial Meanings, Interpretations and Decisions

The parties "size each other up" and make initial judgments as to whom they are
dealing with and what to expect. They also make initial action decisions.

It is important to create space for dialogue at this stage, which will be helpful for
the next stage. Patience, sensitivity, flexibility, attention and leaving room for quick
changes are some of the watchwords for this stage.

As previous chapters showed, initial meanings given to the mineral exploration project by communities covered the entire spectrum from being an environmental and social threat to being an economic opportunity, with associated interpretations of the situation. For communities that saw the project as an opportunity, local general assemblies were the most common reference community that influenced their decisions for action. A number of the communities opposed to mining drew on a wider range of reference communities that included organizations beyond their immediate region such as NGOs and aboriginal and political organizations that predated the arrival of the project. This means that the initial "lay of the land" had a significant effect on subsequent interactions. It may therefore be prudent for host governments to develop policies that help shape the lay of the land well in advance of mineral exploration, and to take the sociological approach described in this book into account in the design of these policies. "Home" government agencies would also benefit from considering such factors, for example, they may affect Export Development Canada's lending decisions.

6.3   Stage 3: Meeting Transactional Needs

At this stage, the parties set the scene for developing a substantive relationship in the next stage. It involves creating the conditions that allow open discussion of substantive issues and the expression of emotions. Solidifying dialogue space would be the first step, followed by trust building taking into account its tactical dimensions: being visible, personalizing encounters, being clear to the other party about your own expectations, and establishing routines. Ideally at this stage, the parties honestly, respectfully and transparently explore each other's identities. They expect to benefit from their effort and want to be fully included in the "flow of things". Progress on this front will likely require a number of iterations through stages 3, 4 and 5. The degree to which the parties' interactions meet transactional needs strongly influences the characteristics of the relationships that develop in the next stage. In most cases, the iterations through the cycle at the heart of the model gradually changed meanings, reference communities and actions, which resulted in strengthening the local communities and in a productive relationship between the company and the communities.

The initial meanings that exist in communities opposed to mining create barriers to meeting transactional needs, especially in the absence of dialogue. In the present study, this occurred to varying degrees in a number of communities. While a number of the six communities in the area of influence of the Cauchari-Olaroz project initially saw the project as a threat, they adjusted the meanings they gave through intense interaction with the other communities which led to a joint decision to establish a dialogue with the company. Most other communities involved in the case studies decided either to work proactively with the company, or not to oppose its activities. With these communities, the process of meeting mutual transactional needs began "in earnest". Building trust and meeting transactional needs are at the heart of the model: they enable all other processes. For example, in the Lindero case, Tolar Grande and Mansfield Minera established trust between them very early on in the life of the project. Both the townspeople and company employees at all levels worked at meeting each other's transactional needs. As a result, the company became fully integrated into the community to the point that community members with general concerns not related to the project were as likely to contact the director of Mansfield Minera as they were to contact the town's mayor, and both saw this as a normal part of everyday life. In some communities, the barriers to meeting transactional needs were insurmountable.

6.4   Stage 4: Building Relationships

Meeting the transactional needs of both parties creates the conditions for developing relationships along with the dimensions trust; respect; communication; mutual understanding; conflict resolution; goal compatibility; balance of power; stability; and productivity. Which of these dimensions will develop most depends on the local situation and the needs and characteristics of the parties. The shades of red and green that fill the cells of Table 5.1 largely reflect the degree to which the parties met transactional needs and the consequent characteristics of the relationships. Using the conflict resolution dimension best illustrates the effect of meeting transactional needs on relationships. Conflicts inevitably occur and, if managed well, can drive necessary change. While various authors see a need for the establishment of formal conflict resolution mechanisms (Onuoha & Barendrecht, 2012; Rees, 2009), explicit conflict resolution mechanisms were not present in any of the case studies described in this book. In the cases that did meet transactional needs, existing interaction mechanisms apparently served multiple functions (that included conflict resolution, joint planning and budgeting, organization of events and carrying out joint projects, which underlines the importance of meeting transactional needs). Personal relationships were very important in all cases.

While relationships are very important in the community engagement section of most CSR guidelines and codes, none pays specific attention to relationships per se. A review of these guidelines and codes through a sociological lens may reveal opportunities for improvement.

The processes and concepts are the same for all cases, but because of differences in content, the cases remain sui generis; community-company relationship characteristics cover a broad range, from strong animosity to friendship; and meeting transactional needs is a precondition for the positive development of relationships and for changing meanings.

6.5   Stage 5: Change Processes

All cases involved changes of meanings and considerable influence of reference communities. Changes of meanings often began with individuals or groups that did not share the meanings of the majority. The change processes were not linear and contingent factors (such as the reputation of the change agents, as was the case in the parish of Los Encuentros; the unity and determination of the women's organization in Chumblín) influenced the convergence of meanings between communities and companies. Meeting transactional needs aided change and involved considerable effort on both sides (comment by an interviewee in San Gerardo—Loma Larga case). In a number of communities, meanings did not significantly change with time.

Interactions with one's fellows adjust meanings and the party that has the most effective and intensive interactions between its members and that can call on the deepest reservoir from which to draw possible meanings has the highest probability of being able to change "in-group meanings". In many cases, this will be the company because the interaction between its employees is much more frequent and intense than between community members. In addition, the relationship with the community is an area of focus for company employees, while the relationship with the company is only one of many relationships for community leaders and members. In many cases, the company has a higher proportion of educated people than does the community and its resources allow it to gain access to much deeper reservoirs of meanings on which to draw—in effect the company can draw on a wider reference community with greater ease. The most prominent reference communities on the community side were general assemblies and community leaders, and the Catholic Church and the Rondas Campesinas in one of the cases not described in this book. While most communities were not consciously aware of this, factors "farther afield" in the wider field of reference mentioned earlier influenced some of their actions. Examples are changes in legislation that have given more power to communities in Peru (related to Free, Prior and Informed Consent) and in Ecuador (the Buen Vivir concept and the 2009 mining law that I mentioned earlier). The concepts presented by a member of one of the six aboriginal communities in the zone of direct influence of the Cauchari-Olaroz project bore similarities to those found in international codes and guidance notes (e.g. e3 Plus), suggesting that these documents were part of her extended reference community, demonstrating the convergence of meanings that had occurred (Calpanchay, 2012).

Reference communities for the companies included their community relations team (reference community for its members and other company staff) and a variety of local actors (e.g. in the Fruta del Norte, Loma Larga, Lindero and Cauchari-Olaroz cases in which the company and the community maintained close links). In addition, country management, international parent company, international social responsibility codes, country legislation, mining associations and home country influences were sources of the norms that informed their choice of options for action.

The meanings, reference communities and relationships observed at the time of the case studies resulted from continuous iterations through stages 3, 4 and 5, and going into the future, this loop will keep feeding changes into the next stages. The above considerations imply that all actor groups benefit from paying close attention to:

- meeting transactional needs; identifying those who hold alternative meanings and are respected in their communities and work with them to try and establish a dialogue with those who hold majority meanings, all the time meeting transactional needs
- applying Gawley's tactical dimensions (visibility, personalized encounters, "showing face", establishing routines)
- maintaining presence (e.g. in the Loma Larga and Fruta del Norte cases the community relations team stayed "on the ground" even during interruptions of exploration activity)
- learn from interactions and adjust as necessary.

The iterative processes described above led to the situations that prevailed at the time of my study and to the relationship characteristics as summarized in Table 5.1.

6.6   Stage 6: Relationship Patterns and Social Structures

In most case studies, the arrival of the exploration project had a significant impact on relationship patterns. With time, relationships and related interactions led to changes in meanings: convergence in many cases, divergence in others such as the community of Victoria del Portete (Loma Larga). At first sight, an exploration project would appear to be just a new node added to the existing networks of relationships. However, the relationship matrices for almost all case studies showed the existence of only a small number of relationship "axes" (entities interacting with a large proportion of all entities and individuals). The major axes were always the exploration company, together with the authorities of the local community and in some cases the Catholic Church and an aboriginal or community organization (e.g. the Shuar Federation in the Fruta del Norte case). The precarious economic circumstances of a large proportion of the population as evident from the socio-economic indicators made it possible for the resources of even a relatively small exploration project to have a significant impact on the local economy. This, together with the economic promise or environmental or social threat associated with a potential mine, could account for the local prominence of the project. Another important factor leading to the strong impact of companies on local relationship patterns was that most actively sought to establish relationships with community entities and individuals.

Interaction processes function better when they are embedded in a social structure as it provides information about which of the factors such as status, role, aspects of culture and "system ecology" are most important and thereby it gives clues as to how best to go about meeting transactional needs. This tends to reinforce the social structures that embed the interactions through the identity verification processes mentioned earlier (Turner, 2011). This may be one of the reasons for the stability of the relationships in projects that integrated their CSR strategies with local governance structures such as development and land use plans and parish councils (Loma Larga and Fruta del Norte) or that maintained tight linkages and cooperation with local general assemblies, town councils and provincial authorities (Cauchari-Olaroz, Lindero). The tactical dimensions of trust development, especially consistency, presence and establishing routines also have important implications. In communities not opposed to mining they can strengthen the social structure of communities, while in conflictive situations a negative tit-for-tat routine can develop that deepens the rift between the parties and leads to a descending conflict spiral.

The social structures mentioned above acquired most definition in the cases that developed the greatest clarity about who should be legitimately involved in the relationship and about the matters at stake. This happened in the cases that had established clearly articulated mutual roles and expectations together with mechanisms to meet these expectations: Loma Larga (for the communities no longer opposed to mining), Cauchari-Olaroz, Fruta del Norte, and Lindero. For example (as was mentioned in Chap. 5), the collaboration between the communities and the company strengthened community organization. Actors in these cases met transactional needs and thereby recursively strengthened the social structures that formed, which in turn led to greater stability. Inspection of the "stability" column in Table 5.1 shows a general agreement with the predicted trend. This stage of the model has important societal implications: the establishment of the new social structures can change local governance and politics, for example, increasing the legitimacy of the parish councils in Los Encuentros (Fruta del Norte case) and Chumblín (Loma Larga case) led to improved governance. The next section considers a number of additional factors that feed into the outcomes. A discussion of stage 7 will follow.

6.6.1   Additional Factors Feeding into the Core Processes

Interactionist processes drive the overall dynamics, but a number of additional factors feed into the outcomes. They include community characteristics (cohesion, leadership, capacity and prior history), company characteristics (management style, culture, skill set, resources and CSR strategy), time and contextual factors such as socio-economic conditions, history and government requirements. Among the company characteristics, the approach to social responsibility has a major impact on perceptions of benefits and harms. Both community and company characteristics can influence and be influenced by the core processes.

6.6.1.1   Community Characteristics

As the course of social events involves community actors, community characteristics inevitably play a role. Important factors in this respect include internal level of trust, governance (leadership, organization and cohesion, skills and capacity) and personalities.

Internal level of trust in the population is an important independent variable because it affects the functioning of the social networks that transmit and change meanings and through which reference communities provide norms. It also affects the way in which communities organize themselves. For example, the overall low level of trust between and within communities in the Lurín valley affected their ability jointly to take full advantage of the presence of the Palma project to support regional development projects. On the other hand, the same factor reduced the ability of those opposed to mining to mount coordinated resistance. Local governance as expressed through community leadership, organization and cohesion can lead to a constructive dialogue and positive outcomes for both parties (in terms of both the social course of events and economic and development benefits) as was the case in the parish of San Gerardo (Loma Larga case) and in the communities surrounding the Cauchari-Olaroz project. The same characteristics can also have a negative effect on the course of social events such as in the Río Blanco case (not described in this book), where they led to an intensification of the conflict and elevated it to a regional and national level (Boon, 2015).

In view of the importance of personal relationships, actor personality characteristics can significantly influence the course of social events. For example in the Lindero case, the personality of the mayor of San Antonio de los Cobres made it impossible for any of the companies working in the area to undertake social projects in this town. In the Palma case, the lack of assertiveness of a local authority slowed down decision-making and led to seemingly endless general assemblies:

> ...It is as though he is afraid and he feels he needs to seek approval for the smallest decisions [company personnel]...

6.6.1.2 Company Characteristics

The chapter "Sociology for mineral exploration" in the forthcoming book "Geoethics: Status and Future" that will be published by The Geological Society of London in 2020 links mineral exploration company characteristics such as management style, culture, skill sets, resources and social responsibility strategy to the core concepts of sociological theory and provides a framework for action. It presents easy to use tools for managing relationships within the company and with the communities surrounding the project. This forthcoming chapter draws on the case studies that also formed the basis for the present book.

6.6.1.3 Contextual Influences

It is not possible to capture project-specific contextual factors in a generalized model. They affect the course of social events and the perceptions of the present and potential future benefits and harms of any project. These need identification and consideration on a case-by-case basis. The section below describes a few of the contextual factors that played a role in the present group of case studies.

In the Fruta del Norte and Loma Larga cases the government requirement of development and land use plans for strategic zones, together with the involvement of the national government greatly facilitated the design and execution of joint company-community projects, and in the Cauchari-Olaroz case full community control over the land was an important factor in establishing power equilibrium. The concepts of community engagement and community development in the CSR frameworks employed by industry evolved over the past decade and affected company approaches (International Council on Mining and Metals, 2010; Prospectors and Developers Association of Canada, 2010; The Mining Association of Canada, 2014). Another important factor is that of aboriginal rights and culture that played an important part in the Cauchari-Olaroz case.

In summary, contextual factors modulate the generalized model. Such factors are specific to each situation. In the present study, several of the contextual factors that interviewees mentioned related to government policies. This emphasizes the important role played by governments. Their decisions always have consequences, even if they decide to do nothing.

6.6.1.4   Time

The time scale on which communities operate is very different from that on which companies operate. The former have a long-term orientation, the latter a short-term orientation. Especially in remote regions, community rhythms tie to natural cycles such as birth and dying, climate, crops and animals. Commodity prices, stock market movements, investor expectations and technical requirements drive the rhythms of exploration companies. The difference in time scales is a major source of misunderstandings and conflicts. Building relationships takes time, often more time than is available to the company. There are legion examples of projects going off the rails because companies made shortcuts to save time, for example in the Esquel case that I mentioned earlier (WRI: Development Without Conflict, 2007). It is possible to make more speed with less haste and building up the project slowly and cautiously over a long period as was demonstrated by the Lindero project, which is quite unusual in the industry. While this required finding patient investors and ongoing thoughtful investor relations management, it helped it to integrate into the community, establish excellent relationships in the region and influence the approaches taken by other companies working in the area.

6.7   Stage 7: The Course of Social Events and Perceived Present and Future Benefits and Harms

The examples that I used to illustrate the various features of the generalized model amply demonstrated how the proposed processes explain the course of social events. On the positive side, they included full integration of the company into the community; communities uniting to work together with the company; functioning informal conflict resolution approaches; strengthening of local social structures and establishment of new social structures; many examples of personal relationships triggering major improvements; and decreases in internal division of communities. The same interactionist processes that led to the convergence of meanings between communities and companies on the positive side led to the convergence of the negative meanings community actors gave to the exploration project, which intensified conflict in a number of communities. The social outcomes of the iterative interactionist relationships cycle were a precondition for the realization of additional practical benefits through cooperation between the parties. Perceived present and future community benefits mentioned by interviewees in addition to the social aspects mentioned above fall into the categories economic, health and safety, capacity building and education, sports, infrastructure development and other.

Economic benefits included employment (in some cases expectations were unrealistically high and needed to be managed) and activities and business development generated by the project such as local purchases by employees, provision of services to the project and a hotel to provide accommodation for external visitors to the project. In addition, it included support for the establishment of businesses and business associations (for example a fisheries cooperative) and establishment of cooperatives (for example in organic produce). Benefits in health and safety included the establishment of a medical post and attracting a medical doctor; CSR initiatives such as cancer and eyesight screening, vaccination of cattle; medical supplies; and special attention to and care of the weak and elderly. In capacity building and education, there were practical courses in bakery, carpentry, plumbing, electricity, masonry, agriculture; organizational development; provision of a teacher and a building for early childhood education; and in cooperation with the ministry of education, high-school level education for a significant proportion of the population. Support for sports activities included regional soccer championships; a soccer school and a successful local soccer team; and sports facilities. Infrastructure development projects included road construction and improvement; improvement of drinking water supply; and construction of religious buildings. Miscellaneous other reported contributions or outcomes were assistance with the resolution of land ownership issues; reduced emigration of breadwinners and return of emigrants to their families (mentioned by interviewees in more than one country); improvement of personal lives; and good personal relationships.

The companies also benefited from the interactionist processes of the model. First, they moved towards a company culture that could meet the needs of the situation. Second, they experienced additional practical benefits, the most important of which was the ability to conduct their work without stoppages, in a tranquil environment. Most community interviewees felt that the real benefits for the company would only come once mining operations would start in earnest. The companies that had community development as one of their explicit goals were pleased to see progress. It is important to realize that the joint effort of *both* parties led to the achievement of the benefits described above.

While the communities that had established positive relations with the project did not perceive actual harms, some unease about potential future environmental damage remained. Various community members expressed a nuanced opinion and recognized that they too contaminated the rivers. In addition, while there was satisfaction with the present relationship, some wondered about how it would change when mining operations began or when there was a change in management or company ownership. For example, Goldrock bought Mansfield Minera after the present study was completed, and Goldrock decided to keep the name "Mansfield Minera" because it was cognizant of the importance of the meaning this name has for the local communities.

The communities that did not establish a positive relation with the mineral exploration project perceived much present and future harm. Their list of concerns included severe environmental damage (with concerns about water quality topping the list), social upheaval and destruction of the social fabric. In addition, they were suspicious about the intentions of the company ("…they will take the resource and leave us with the mess…"), feared loss of culture through Westernization. They saw conflicting development models; co-optation and corruption of authorities and institutions; and government criminalization of protests. There had been few real encounters between these groups and the "other camp", which precluded the process of meeting transactional needs and the consequent creation of a dialogue space.

My study confirmed that relationships do indeed affect the course of social events and the perceptions of the present and future harm and explained the processes through which this happens. While most people in mineral exploration agree that relationships play an important role in their field and while most codes and guidelines emphasize the importance of establishing good relationships, there has been little in-depth study of the innards of relationship building. In contrast, there are mountains of the literature on the hydrothermal or other earth processes that lead to the formation of ore deposits and any exploration geologist or geophysicist is familiar with a slew of ore deposit models. The present study has begun to rectify that imbalance by developing a model based on a small number of relatively simple concepts: meanings, interpretations, reference communities and related change processes. Geologists effortlessly wrap their creative minds around complex models of ore formation that plays out over millions of years and that involve many hundreds of cubic kilometres of rock and structures that are only partly visible. I hope that they will see fit to apply their considerable talents to develop the relationship process model presented in this book.

References

Boon, J. (2015). *Corporate social responsibility, relationships and the course of events in mineral exploration—An exploratory study* (Unpublished Ph.D. thesis). Carleton University, Ottawa. https://curve.carleton.ca/system/files/etd/6c6598d4-c436-409e-9ba1-40dea2d37d2c/etd_pdf/7b39ca613ff7e7e2df52ed82580e3974/boon-corporatesocialresponsibilityrelationships.pdf.

Calpanchay, R. (2012). Comunidad aborigen de Puesto Sey "Termas del Tuzgle". Retrieved February 10, 2014 from www.olami.org.ar.

International Council on Mining and Metals. (2010). Sustainable development framework. Retrieved December 28, 2010 from http://www.icmm.com/our-work/sustainable-development-framework.

Onuoha, A., & Barendrecht, M. (2012). *Issues between company and community: Towards terms of reference for CSR-conflict management systems.* The Hague: Hague Institute for the Internationalization of Law—Innovating Justice.

Prospectors and Developers Association of Canada. (2010). e3PLUS—A framework for responsible exploration. Retrieved December 28, 2010 from http://www.pdac.ca/e3plus/index.aspx.

Rees, C. (2009). *Report of international roundtable on conflict management and corporate culture in the mining industry* (No. 37). Cambridge, Massachusetts: Corporate Social Responsibility Initiative, Harvard Kennedy School, Harvard University.

The Mining Association of Canada. (2014). Towards sustainable mining. Retrieved August 22, 2014 from http://mining.ca/towards-sustainable-mining.

Turner, J. H. (2011). Extending the symbolic interactionist theory of interaction processes: A conceptual outline. *Symbolic Interaction, 34*(3), 330–339.

WRI: Development Without Conflict. (2007). Case 2. Esquel gold project, Argentina. Retrieved February 13, 2014 from http://www.sarpn.org/documents/d0002569/8-WRI_Community_dev_May2007.pdf.

Chapter 7
Implications and Conclusions

The description of the model included a number of implicit and explicit suggestions for action by various actors. The sections that follow further elaborate on how to use the model to frame the approach communities, companies, governments and organizations take in dealing with situations and with each other. The first section discusses approaches that apply to all actor groups, and subsequent sections elaborate aspects specific to particular actor groups.

7.1 Implications for All Actors

7.1.1 Reference Communities

Knowledge of reference communities provides insight into community and company decision-making processes. While the reference community concept and methods for identifying and characterizing these communities need further study, it is possible to make a first approximation that can serve as an initial reference frame for understanding the situation and identify potential avenues for change. This applies to both the communities and the company. A preliminary map of reference communities, the values they espouse and the meanings they give would be useful. Areas of overlap between reference communities would be of special interest, as would identification of intermediaries: persons respected by more than one reference community. The intermediaries can be key resources in bringing about convergence of meanings. Companies are possibly better equipped for these tasks than are communities, but as the Cauchari-Olaroz and San Gerardo (Loma Larga) examples showed quite capable communities do exist and it is very important not to make assumptions about the capacities of communities. Tracing the personal networks within reference communities would help define their extent and possible overlap with others, as well as identify "outliers", potential nuclei of change.

© The Author(s), under exclusive license to Springer Nature Switzerland AG 2020   117
J. Boon, *Relationships and the Course of Social Events During Mineral Exploration*,
SpringerBriefs in Geoethics, https://doi.org/10.1007/978-3-030-37926-1_7

7.1.2 Meanings

Meanings given to people and things by both community members and company personnel determine their interpretation of situations and influence their actions. In communities, many of these meanings are implicitly discussed, formed and changed in general assemblies. Outliers are crucial to help avoid communities locking in too soon into particular sets of meanings and options for change. While some community leaders may be aware of this, many probably are not and there is a role for higher-level authorities and NGOs to build capacity in this area. On the company side, this responsibility falls to management. Dialogue between community and company is essential to achieving convergence of meanings: each side needs to learn from the other.

7.1.3 Self-analysis

A self-analysis will provide valuable clues as to weaknesses in addressing transactional needs, knowledge of the meanings a group gives to people and things (especially the community) and of its reference communities. Conducting an early self-analysis of a group's internal relationship indicators, meanings and reference communities can be extremely useful preparation for its interaction with new neighbours. This will allow it to put its own house in order before engaging with other actors, while at the same time developing familiarity with the tools that it may use for external engagement. This will enhance its ability to meet transactional needs. This will require much meaningful discussion within the group. To understand internal social dynamics and ensure that all members of the group are fully included in the debate it may be useful to conduct an internal network analysis. Such analysis will contribute to the identification of weaknesses. For example, it is well known that organizations with very tight internal networks may find it difficult to interact flexibly with other groups. Groups with a somewhat looser network are better at interacting with other groups (Degenne & Forsé, 1999). The approach to situations in which communities are mostly in favour of mining is quite different from that in situations where communities are mostly opposed to mining. In those mostly in favour of mining, it may be relatively straightforward to design and implement the strategies implicit in the model with a reasonable probability of success: the spectrum of meanings will converge and with time, there will be overlap between the meanings and reference communities of all actor groups. It is much more difficult to work towards meeting transactional needs between different groups of actors in situations where communities are strongly opposed to mining, and in some cases, it may be impossible to establish initial interactions. Under such circumstances, outliers (i.e. people giving meanings that are significantly different) and respected intermediaries can play an important role. However, all parties involved in trying to bring about change need to be patient and move with great caution and should accept the possibility that it is not possible to establish productive relationships.

A number of these suggestions are already implicit or explicit parts of various existing guidelines and models. However, the present study integrates them into a coherent sociological perspective and explicitly builds on the underlying mechanisms. Concepts that are novel in comparison with existing guidelines and models are meanings, reference communities, and meeting transactional needs. The coherent sociological framework provides greater depth and hopefully further research and refinement will increase its usefulness and spur the development of innovative approaches.

7.1.4 Creating Dialogue

Meeting transactional needs is a conditio sine qua non for most processes of the model. After an official introductory meeting, a series of semi-formal get-togethers can create the psychological space within which one can begin to meet transactional needs. These get-togethers serve to get to know each other and, while sharing information about the project as part of the conversation, this is not their main purpose. Informal conversations with people in public places will also help actors to get to know each other. The transactional needs (identity confirmation, benefits from the encounter, being part of the flow, trust and absence of hidden agendas) should guide the interaction processes, and from a practical perspective should use the tactical dimensions proposed by Gawley (visibility; personalized encounters; "showing face"; and establishing routines). Almost all suggested approaches can and possibly should be undertaken by each group of actors (albeit that they would need to be adapted, sometimes in significant ways) to get dialogue under way. Creating a dialogue is a prerequisite to achieving the relationship goals implicit in the relationship indicators: trust, respect, communication, mutual understanding, conflict resolution, goal compatibility, balance of power, focus, frequency, stability and productivity. Many of the actors involved in or touched by mineral exploration and mining frequently use the word "dialogue" and most think they know what it is and feel that they can engage in dialogue. However, there is a wide range of understandings of the term including "exchange of views between parties", "tool for resolving conflicts", a "mechanism to influence", "spontaneous open conversation", "wasting time talking" and others. Many have heard of it or talk about it, but few have practised it. As the Dutch proverb says: "they heard the bell ring, but do not know where to find the clapper". Alfredo Bambarén, in his prologue to the book Diálogos que Transforman (Dialogues that Transform) (López Follegatti, 2015) gives dialogue a deeper meaning. He says "Dialogue is not only a matter of the rational sphere, it is a way to arrive at those feelings and perceptions that remain hidden in the unconscious…It involves an attitude that listens without judgment and without resistance and that requires a conscious and deliberate practice of patience, humility, tolerance and positive mind. It is a different way of looking". He quotes Heidegger who said, "It is not us who maintain conversations. Conversations maintain us".

In the introduction to his book, López Follegatti says, "the idea that I want to develop in this book is that there are dialogues that are not spontaneous, but that are born with a clear objective of exchange and that can be learned and practised when conditions are appropriate". Meeting transactional needs helps create appropriate conditions. In the first chapter, López Follegatti clearly shows that dialogue includes a philosophy of life that includes humility, empathy, learning, patience, tolerance and much more, and he links it to relationships. Living this philosophy can lead to personal, institutional and social transformations, with great benefits for individuals, communities, companies and governments at all levels and in all areas. Dialogue can affect governability, and the chapter analyses the functioning of dialogue in social networks, families and schools. The second chapter gives actual examples and learnings linked to dialogue spaces, transformation of participants, the role of organizers and facilitators, consensus, peace and non-violence and the necessity to have dialogues about development. The third chapter makes a series of practical recommendations for the venue, the facilitator, companies, NGOs, communities and the government. The fourth chapter consists of a series of dialogue practice exercises. Rather than providing further details on the practicalities of dialogue, I suggest that the reader consult the book by López Follegatti that is available online (López Follegatti, 2015). In addition, local members of organizations such as the Grupo de Diálogo Latinoamericano would be pleased to provide advice and support (Grupo de Diálogo Latinoamericano, 2013).

Dialogue is a nonlinear process and bumps in the road will undoubtedly occur. The philosophy of life outlined by López Follegatti helps overcome these bumps. In a number of the case studies initially only a few individuals or groups were involved in the dialogue, but through the interaction of these individuals and groups with their fellows, the meanings the latter gave to the exploration project changed and dialogue with a wider group became possible. All actors should be on the lookout for changes of this type and adjust their processes accordingly. In some circumstances, it is not possible to achieve conditions for establishing a dialogue, in which case it may be necessary for the parties to disengage and explore alternative options.

7.1.5   Relationship Development

It is important to understand both the patterns of relationships and their characteristics. It would be wise to "prepare the ground" well in advance. For a company, this could mean to begin "exploration for relationships" in areas of high mineral potential, well before exploration for minerals is to start and to draft a relationship matrix. For communities and local governments, this would mean acquainting themselves with the companies interested in the area, which could take many forms and may require assistance from higher levels of government. The latter could set up partly standard-ized approaches for assistance to communities. "Staking rushes" can pose a particular challenge, as they may generate more requests for meetings and negotiations than communities can handle. The construction of bilateral relationship matrices based on initial survey information can visually capture the findings of these "explorations for relationships". Graduate students of universities in the area could conduct the surveys. These matrices provide a useful baseline to help set priorities and as a start-ing point for relationship development. Through ongoing dialogue conducted along the lines outlined above, relationships will continue to develop from the initial pat-terns. Where possible, the parties should regularly characterize their internal and external relationships using the indicators to track dynamics and make changes as necessary and possible. The resulting relationship patterns can be strengthened by building routine processes around them, such as regular meetings between company management and town council; tripartite meetings between company management, town council and community organizations or higher-level authorities. Such routine processes lead to new social structures that in turn help stabilize relationships and create a context within which the parties can work towards the achievement of their respective goals. As was mentioned in the previous chapter, the iterative loop at the centre of the model will continue as long as there is a project, and its processes need continuous attention.

7.2 Aspects Specific to Particular Actor Groups

7.2.1 Home Government

Home governments' policies can set the tone for the behaviour and approaches of companies with headquarters in their jurisdictions. For example, the Government of Canada recently announced its "Enhanced Corporate Social Responsibility Strat-egy to Strengthen Canada's Extractive Sector Abroad" (Government of Canada, 2014). It expresses the Government of Canada's "…expectation that Canadian com-panies will promote Canadian values and operate abroad with the highest ethical standards…". The strategy will promote greater use of existing CSR guidelines and codes by the industry. In April 2019, the Government appointed the first Cana-dian Ombudsperson for Responsible Enterprise (CORE)—the first of its kind in the world. The Ombudsperson has the mandate to review alleged human rights abuses arising from a Canadian company's operations abroad, make recommendations, mon-itor those recommendations, recommend trade measures for companies that do not co-operate in good faith and report publicly throughout the process. The CORE's scope will focus on the mining, oil and gas and garment sectors. The Ombudsper-son is one of Canada's two voluntary dispute resolution mechanisms, complement-ing Canada's National Contact Point for the OECD Guidelines for Multinational Enterprises (NCP) (Global Affairs Canada, 2019).

The Government uses a carrot and stick approach: it provides enhanced Government of Canada economic diplomacy for companies that align with CSR guidelines. It withdraws such support from companies that do not embody CSR best practices and refuse to participate in the Ombudsperson's or National Contact Point (under the OECD guidelines for multinational enterprises) dispute resolution processes (OECD, 2011). The threat of withdrawal of support for companies that do not meet expectations implies that the Government needs to develop a mechanism for identifying those companies. For example, Trade Commissioners and staff need to build networks and local partnerships with communities and need to be equipped to detect issues early on and contribute to their resolution before they escalate. The model described in the previous chapter provides a context for building networks and partnerships with local communities, as well as for detecting issues early on and suggesting possible solutions. Awareness of the processes involved in the stages $3 \Rightarrow 4 \Rightarrow 5 \Rightarrow 3$ loop will be especially helpful in this respect. The model supports the Government of Canada strategy: "… creating dialogue spaces and venues for bridge-building between companies, communities, and other interest groups… stepping up efforts to support engagement between companies and communities, including at the exploration stage…" and promoting "…meaningful and regular dialogue between companies, local communities, civil society and host country governments at all levels…".

7.2.2   Host Government

Host governments are important actors that, like all other actors, carry their share of social responsibility linked to mineral exploration projects. Its policies, regulations and actions (or lack thereof) set the context within which social events develop and benefits or harms occur. Its ongoing participation in the processes surrounding mineral exploration projects will help ensure that the new social structures that form will be productive and that there is continuity. Host governments and their agencies can use the model to help analyse existing situations in their jurisdictions and decide where government participation would be most beneficial. Relationship matrices and relationship indicators will be useful tools for governments and their agencies in this respect, while at the same time giving them insight into their own performance in the "relationship space". Credibly facilitating the creation of dialogue spaces, where appropriate in cooperation with host governments, can have a significant impact on social events. For example, the government of Peru has significantly moved in this direction through the creation of the Oficina Nacional de Diálogo y Sostenibilidad (National Dialogue and Sustainability Office).

7.2.3   *Industry*

Social Impact Analysis and Stakeholder Analysis (also called Stakeholder Engagement and Representation) have become standard tools that provide a wealth of useful information and suggestions for organizational approaches that can help achieve the parties' goals (Boutilier, Henisz, & Zelner, 2013; Franks, 2012; World Bank, 2012). They view "engagement" as an important separate step in the overall process but do not pay detailed attention to the underlying processes. In addition, they risk taking a detached view that sees situations as needing to be "managed" rather than "lived".

Social impact assessment and stakeholder analysis can incorporate the interactionist model developed in this book at the data collection phase by including questions on meanings, reference communities and relationship characteristics in the field surveys. Below I propose some idealized approaches that can influence the course of social events and the associated perceptions of the present and future benefits and harms. Each particular situation and group of actors would need to choose the approaches most appropriate to its particularities and adapt them as needed. I did not list the approaches in the order in which to apply them—there will be much iteration and different activities could go on at the same time, according to the demands of the situation. In addition, as all human interaction processes take time and are potentially full of surprises, significant time should be set aside and planning should allow for unexpected events.

My study suggests that adopting a flexible, relaxed management style that trusts, respects and involves employees will also contribute to external relational success, as will instilling and living a strong value system in the organization. Thorough diffusion of this approach throughout the company is important.

While making the business case for CSR is de rigueur in industry practices, focusing on it too much risks instrumentalizing the approach and overlooking the importance of relationships and the all-important processes in stages 3, 4 and 5 of the model. Various industry codes mention "gaining a social licence to operate" as one of their objectives. As I mentioned earlier, the "social licence" concept has come under fire, and for good reason. Some practitioners such as Ian Thomson and Bob Boutilier, make the point that "social licence to operate" is a process, not a transaction (Boutilier & Thomson, 2011). The word "licence" suggests an instrumental, purely transactional mechanism, and many in the industry treat it as such. The relationship model shows that the mineral exploration world does not really work this way and that treating relationships the same way one would treat, say, a fishing licence, risks inviting trouble.

The model provides a scientific basis for industry tools such as the excellent community engagement toolkit of e3 Plus. Some of the tools I used in my study (such as the relationship matrix, the relationship indicators) can usefully be added to the tools provided by e3 Plus and other guides, and the sample questions provided in the engagement toolkit could be tweaked or expanded to obtain information about meanings and reference communities. The latter could lead to useful maps.

In summary, the model provides a cohesive theoretical framework for most aspects of the community engagement module of e3 Plus, the leading set of guidelines for the mineral exploration industry, and it lays a basis for further development.

7.2.4 Communities

Some communities may find it difficult to follow up on the general suggestions made in an earlier section because they lack resources or skills. Avenues open to them include exploring assistance from local universities and networks such as members of the Grupo de Diálogo Latinoamericano. Host governments could play an important coordinating role in this respect.

7.3 Conclusions

I conducted nine case studies of exploration projects to find out how relationships within and between the actor groups involved in mineral exploration projects influence both the course of social events and perceived the present and future benefits and harms. I interpreted the results using the framework of the sociological theory of symbolic interactionism, which led to the formulation of a generalized model that links relationships and interactionist processes to the course of social events and to perceptions of present and future benefits and harms. The model's seven stages are arrival of the exploration project; initial meanings, interpretations and decisions; meeting transactional needs and applying tactical dimensions of building trust; building relationships; change processes; relationship patterns and social structures; social events and perceived benefits and harms. Stages 3, 4 and 5 form a continuous iterative loop. Additional factors include community and company characteristics, time, and contextual influences that feed into the processes at various stages. Application of the model to social responsibility approaches increases their transformational power. While the cases were sui generis, the underlying processes were similar. Interactionist concepts key to the interpretation of the case studies were meanings, reference communities, relationships, transactional needs, and dimensions of trust building. The indicators' trust, respect, communication, mutual understanding, conflict resolution, goal compatibility, balance of power, focus, frequency, stability and productivity characterized relationships. These indicators successfully differentiated the cases in terms of qualitative risk of conflict measures. Various cases displayed transformational characteristics as defined by Bowen, Newenham-Kahindi, and Herremans (2010), for example, Fruta del Norte, Cauchari-Olaroz and Lindero. In addition to providing a framework for analysis, the theory of symbolic interactionism provides a framework for policy development for both governments and industry, especially as it shows ways of incorporating "soft" issues into their considerations, and for communities. The most important conclusion is that the model presents a coherent framework for understanding the relationships between the actors and the underlying processes. It also explains how these influence the course of social events and perceptions of the present and future benefits and harms. Ethics and values play an important part in all processes and the Cape Town Geoethics Statement is directly applicable, especially where it mentions responsible resource development.

References

Boutilier, R., Henisz, W., & Zelner, B. (2013). A systems approach to stakeholder management. In *SRMINING 2013 2nd International Conference on Social Responsibility in Mining,* Santiago, Chile (pp. 45–54).

Boutilier, R., & Thomson, I. (2011). Modeling and measuring the social license to operate: Fruits of a dialog between theory and practice. In *International Mine Management 2011,* Queensland, Australia.

Bowen, F., Newenham-Kahindi, A., & Herremans, I. (2010). When suits meet roots: The antecedents and consequences of community engagement strategy. *Journal of Business Ethics, 95*(2), 297–318.

Degenne, A., & Forsé, M. (1999). *Introducing social networks.* London: Sage Publications.

Franks, D. M. (2012). *Social impact assessment of resource projects.* Crawley, Australia: International Mining for Development Centre.

Global Affairs Canada. (2019). Canada's Ombudsperson for Responsible Enterprise. Retrieved August 6, 2019 from https://www.international.gc.ca/trade-agreements-accords-commerciaux/topics-domaines/other-autre/csr-rse-ombudsperson.aspx?lang=eng.

Government of Canada. (2014). Canada's enhanced corporate social responsibility strategy to strengthen Canada's extractive sector abroad. Retrieved November 15, 2014 from http://www.international.gc.ca/trade-agreements-accords-commerciaux/topics-domaines/other-autre/csr-strat-rse.aspx?lang=eng.

Grupo de Diálogo Latinoamericano. (2013). Minería, democracia y desarrollo sostenible. Retrieved February 7, 2015 from http://dialogolatinoamericano.org/index.html.

López Follegatti, J. L. (2015). *Diálogos que transforman.* Retrieved January 5, 2016 from http://www.slideshare.net/Grupodedialogomineria/libro-diálogos-que-transforman.

OECD. (2011). 2011 update of the OECD guidelines for multinational enterprises. Retrieved July 28, 2012 from http://www.oecd.org/document/33/0,3746,en_2649_34889_44086753_1_1_1_1,00.html.

World Bank. (2012). *Mining community development agreements—Source book.* Washington D.C.: World Bank.